Food Additives and Human Health

Food Additives and Human Health

Xander Clayton

STATES
ACADEMIC PRESS
www.statesacademicpress.com

Food Additives and Human Health
Xander Clayton
ISBN: 978-1-63989-205-1 (Hardback)

STATES
ACADEMIC PRESS

Published by States Academic Press,
109 South 5th Street,
Brooklyn, NY 11249, USA

Cataloging-in-Publication Data

Food additives and human health / Xander Clayton.
p. cm.
Includes bibliographical references and index.
ISBN 978-1-63989-205-1
1. Food additives. 2. Health. 3. Enriched foods. I. Clayton, Xander.
TX553.A3 F66 2022
664.06--dc23

For more information regarding States Academic Press and its products, please visit the publisher's website www.statesacademicpress.com

Table of Contents

	Preface	**VII**
Chapter 1	**Introduction to Food Additives**	**1**
	a. Function of Food Additives	16
	b. Safety Aspects and Allergenicity of Food Additives	17
Chapter 2	**Food Coloring and Flavoring**	**25**
	a. Food Coloring	25
	b. Classification of Colorants	26
	c. Food Flavoring	42
	d. Natural Flavors Obtained by Microbiological Pathway	59
	e. Sweeteners	70
	f. Chemistry of Food Flavors	78
Chapter 3	**Food Emulsifiers and Stabilizers**	**91**
	a. Bioemulsifiers	110
	b. Emulsifying Salts	117
	c. Food Stabilizers	124
	d. Thickening and Gelling Agents	171
Chapter 4	**Nutrients as Food Additives**	**183**
	a. Minerals as Nutritive Additives	183
	b. Vitamins as Nutritional Additives	199
	c. Proteins as Nutritional Additives	207
Chapter 5	**Enzymes as Food Additives**	**221**
	a. Enzymes Application for Fruits and Vegetables Processing	221
	b. Enzymes Application for Wine Processing	224
	c. Enzymes Application in Baking and Meat Industry	228
	Permissions	
	Index	

Preface

Substances which are added to food items to improve or preserve their flavor, freshness, texture or appearance are known as food additives. Vinegar, salt and sugar are some of their prominent examples. Food additives which are added to fulfill a specific purpose are known as direct food additives. Xanthan gum is its prime example, which is extensively used in salad dressings, bakery fillings, chocolate milk, etc. Indirect additives refer to those substances which become a part of the food product in trace amounts. Depending on the function served by the additives, they are also classified into acidity regulators, antifoaming agents, antioxidants, food colorings, emulsifiers, flavor enhancers, humectants, preservatives and stabilizers. This book is a compilation of chapters that discuss the most vital concepts in the field of food additives. As this field is emerging at a fast pace, it will help the readers to better understand its concepts. For all those who are interested in food additives, this book can prove to be an essential guide.

To facilitate a deeper understanding of the contents of this book a short introduction of every chapter is written below:

Chapter 1- Substances added to food in order to enhance its taste, texture, color and appearance are called food additives. A few types of food additives are foaming agents and anti-foaming agents. This chapter briefly introduces the major aspects of food additives, such as its advantages and disadvantages, and its functions.

Chapter 2- Any pigment added to a food item to enhance its color is referred to as food coloring. The ingredients that are added to any food item to improve its taste or flavor are termed as food flavoring. This chapter has been carefully written to provide an easy understanding of food coloring, classification of colorants and food flavoring.

Chapter 3- An additive which facilitates the mixing of two liquids is termed as an emulsifier. Stabilizer is a type additive that helps to preserve the structure of food to which it is added. The topics elaborated in this chapter will help in gaining a better perspective about the common food emulsifiers and food stabilizers.

Chapter 4- The substances that added to food in order to restore the nutrients which are lost during production, for the purpose of correcting dietary deficiencies, are termed as nutritional additives. The concepts elaborated in this chapter will help in gaining a better perspective about the minerals, proteins and vitamins that are added to food as nutritional additives.

Chapter 5- Enzymes are proteins that can act as biological catalysts. They can be used as food additives for fruits and vegetables processing, wine processing, and baking and meat industry. This chapter closely examines these key applications of enzymes as food additives to provide an extensive understanding of the subject.

Finally, I would like to thank the entire team involved in the inception of this book for their valuable time and contribution. This book would not have been possible without their efforts. I would also like to thank my friends and family for their constant support.

Xander Clayton

1

Introduction to Food Additives

Substances added to food in order to enhance its taste, texture, color and appearance are called food additives. A few types of food additives are foaming agents and anti-foaming agents. This chapter briefly introduces the major aspects of food additives, such as its advantages and disadvantages, and its functions.

Mankind has been adding substances to foods for thousands of years to preserve them, improve their flavour and appearance, and for manufacturing convenience foods. Historians tell us that colourings were added to foods as early as 1500 BC. Theophrastus, a Greek philosopher, reported the use of artificial flavourings in foods in the 3rd century BC. Pliny the Elder remarked on the use of animal fodder and chalk in making bread in the 1st century AD. In the 2nd century AD the respected physician, Galen, warned against the adulteration of herbs and spices. Sodium chloride or common salt has been added to foods as a preservative for thousands of years. Spices like nutmeg and cinnamon, used to flavour foods, have been sought after and traded throughout history. Thc trade routes between Asia, the Middle East and Europe flourished as the demand for these additives grew. In Rome wealthy citizens were kidnapped for ransom, not of gold or jewels, but exotic spices. The ancient Greeks burned sulphur over wine casks before sealing to produce the preservative sulphur dioxide which is still used today.

Making convenience foods both varied as well as nutritious and safe, is now being done using modern technology. This technology includes food additives that have been proven both useful as well as safe through long use and stringent testing. Food additives are substances added to foods to preserve flavor or enhance its taste and appearance. Since prehistoric times, chemicals have been added to foods to perform special functions for example, preserving food by pickling (with vinegar), salting, as with bacon, preserving sweets or using sulfur dioxide as in some wines. Although basic foods contain no additives, as foods are processed for conversion into a variety of products, an increasing number of additives are generally used. Technological advances in food processing have increased the variety and use of these additives. Today, more than 2500 different additives are intentionally added to foods to produce a desired effect. The use of these additives is a well-accepted practice but is not without controversy.

Food additives are defined in various ways.

General

Food additives may be defined as chemical substances which are deliberately added to foods, in known and regulated quantities, for the purpose of assisting in the processing of foods, preservation of foods or in improving the flavor and texture or appearance of foods. Food additives are in general not a primary (major) constituent of food and are usually used at a very small quantity.

Food Additive as per National Academy of Science

Food additives are defined as those chemicals which are incorporated in the food stuffs, either directly or indirectly during processing.

Food Additive as per Food and Agricultural Organization / World Health Organization

"Food additives have been defined as non-nutritive substances which are added either intentionally to food, generally in small quantities to improve their appearance, flavor, texture or other secondary properties or find their way otherwise, into food during handling, processing or distribution." In more recent wordings the term improvement is not included, and the remaining presence of the compound, or reaction products from that compound, is included in the definition.

Many chemical compounds are extracted from foods and added to other foods. In this case they are considered as additives. For example:

- Meat extract from meat.
- Cod liver oil from fish.
- Juice from fruits.
- Starch from grains.

Food Additive as per Codex Alimentarius Commission

"Food additives means any substance not normally consumed as a food by itself and not normally used as a typical ingredient of the food, whether or not it has nutritive value, the intentional addition of which to food for technological (including organoleptic) purpose in the manufacture, processing, preparation, treatment, packing ,transport or handling such food results or may be reasonably expected to result (directly or indirectly) in it or its by-products becoming a component or otherwise affecting the characteristics of such foods."

The term does not include contaminants, insect fragments, hairs and other extraneous material added to food for maintaining or improving nutritional availabilities.

Food Additive as per International Dairy Federation

"Food additives are the chemical substances that might be added to milk and dairy products for any purpose."

Nutritional Additives

Nutritional additives are used for the purpose of restoring nutrients lost or degraded during production, fortifying or enriching certain foods in order to correct dietary deficiencies, or adding nutrients to food substitutes. The fortification of foods began in 1924 when iodine was added to table salt for the prevention of goitre. Vitamins are commonly added to many foods in order to enrich their nutritional value. For example, vitamins A and D are added to dairy and cereal products, several of the B vitamins are added to flour, cereals, baked goods, and pasta, and vitamin C is added to fruit beverages, cereals, dairy products, and confectioneries. Other nutritional additives include the essential fatty acid linoleic acid, minerals such as calcium and iron, and dietary fibre.

Processing Agents

A number of agents are added to foods in order to aid in processing or to maintain the desired consistency of the product.

Processing additives and their uses		
Function	Typical chemical agent	Typical product
Anticaking	Sodium aluminosilicate	Salt
Bleaching	Benzoyl peroxide	Flour
Chelating	Ethylenediaminetetraacetic acid (EDTA)	Dressings, mayonnaise, sauces, dried bananas
Clarifying	Bentonite, proteins	Fruit juices, wines
Conditioning	Potassium bromate	Flour
Emulsifying	Lecithin	Ice cream, mayonnaise, bakery products
Leavening	Yeast, baking powder, baking soda	Bakery products
Moisture control (humectants)	Glycerol	Marshmallows, soft candies, chewing gum
pH control	Citric acid, lactic acid	Certain cheeses, confections, jams and jellies
Stabilizing and thickening	Pectin, gelatin, carrageenan, gums (arabic, guar, locust bean)	Dressings, frozen desserts, confections, pudding mixes, jams and jellies

Emulsifiers are used to maintain a uniform dispersion of one liquid in another, such as oil in water. The basic structure of an emulsifying agent includes a hydrophobic portion, usually a long-chain fatty acid, and a hydrophilic portion that may be either charged or uncharged. The hydrophobic portion of the emulsifier dissolves in the oil phase, and the hydrophilic portion dissolves in the aqueous phase, forming a dispersion of small oil droplets. Emulsifiers thus form and stabilize oil-in-water emulsions (e.g., mayonnaise), uniformly disperse oil-soluble flavour compounds throughout a product, prevent large ice crystal formation in frozen products (e.g., ice cream), and improve the volume, uniformity, and fineness of baked products.

Stabilizers and thickeners have many functions in foods. Most stabilizing and thickening agents are polysaccharides, such as starches or gums, or proteins, such as gelatin. The primary function of these compounds is to act as thickening or gelling agents that increase the viscosity of the final product. These agents stabilize emulsions, either by adsorbing to the outer surface of oil droplets or by increasing the viscosity of the water phase. Thus, they prevent the coalescence of the oil droplets, promoting the separation of the oil phase from the aqueous phase (i.e., creaming). The formation and stabilization of foam in a food product occurs by a similar mechanism, except that the oil phase is replaced by a gas phase. The compounds also act to inhibit the formation of ice or sugar crystals in foods and can be used to encapsulate flavour compounds.

Gum arabic from Acacia species. Gum is used as a stabilizing or thickening agent in food to increase the viscosity of the final product.

Chelating, or sequestering, agents protect food products from many enzymatic reactions that promote deterioration during processing and storage. These agents bind to many of the minerals that are present in food (e.g., calcium and magnesium) and are required as cofactors for the activity of certain enzymes.

Preservatives

Food preservatives are classified into two main groups: antioxidants and antimicrobials. Antioxidants are compounds that delay or prevent the deterioration of foods by oxidative mechanisms. Antimicrobial agents inhibit the growth of spoilage and pathogenic microorganisms in food.

Food preservatives	
Chemical agent	Mechanism of action
Antioxidants	
Ascorbic acid	Oxygen scavenger
Butylated hydroxyanisole (BHA)	Free radical scavenger
Butylated hydroxytoluene (BHT)	Free radical scavenger
Citric acid	Enzyme inhibitor/metal chelator
Sulphites	Enzyme inhibitor/oxygen scavenger
Tertiary butylhydroquinone (TBHQ)	Free radical scavenger
Tocopherols	Free radical scavenger
Antimicrobials	
Acetic acid	Disrupts cell membrane function (bacteria, yeasts, some molds)
Benzoic acid	Disrupts cell membrane function/inhibits enzymes (molds, yeasts, some bacteria
Natamycin	Binds sterol groups in fungal cell membrane (molds, yeasts)
Nisin	Disrupts cell membrane function (gram-positive bacteria, lactic acid-producing bacteria)
Nitrates, nitrites	Inhibits enzymes/disrupts cell membrane function (bacteria, primarily Clostridium botulinum)
Propionic acid	Disrupts cell membrane function (molds, some bacteria)

Sorbic acid	Disrupts cell membrane function/inhibits enzymes/inhibits bacterial spore germination (yeasts, molds, some bacteria)
Sulfites and sulfur dioxide	Inhibits enzymes/forms addition compounds (bacteria, yeasts, molds)

Antioxidants

The oxidation of food products involves the addition of an oxygen atom to or the removal of a hydrogen atom from the different chemical molecules found in food. Two principal types of oxidation that contribute to food deterioration are autoxidation of unsaturated fatty acids (i.e., those containing one or more double bonds between the carbon atoms of the hydrocarbon chain) and enzyme-catalyzed oxidation.

The autoxidation of unsaturated fatty acids involves a reaction between the carbon-carbon double bonds and molecular oxygen (O_2). The products of autoxidation, called free radicals, are highly reactive, producing compounds that cause the off-flavours and off-odours characteristic of oxidative rancidity. Antioxidants that react with the free radicals (called free radical scavengers) can slow the rate of autoxidation. These antioxidants include the naturally occurring tocopherols (vitamin E derivatives) and the synthetic compounds butylated hydroxyanisole (BHA), butylated hydroxytoluene (BHT), and tertiary butylhydroquinone (TBHQ).

Specific enzymes may also carry out the oxidation of many food molecules. The products of these oxidation reactions may lead to quality changes in the food. For example, enzymes called phenolases catalyze the oxidation of certain molecules (e.g., the amino acid tyrosine) when fruits and vegetables, such as apples, bananas, and potatoes, are cut or bruised. The product of these oxidation reactions, collectively known as enzymatic browning, is a dark pigment called melanin. Antioxidants that inhibit enzyme-catalyzed oxidation include agents that bind free oxygen (i.e., reducing agents), such as ascorbic acid (vitamin C), and agents that inactivate the enzymes, such as citric acid and sulfites.

Antimicrobials

Antimicrobials are most often used with other preservation techniques, such as refrigeration, in order to inhibit the growth of spoilage and pathogenic microorganisms. Sodium chloride (NaCl), or common salt, is probably the oldest known antimicrobial agent. Organic acids, including acetic, benzoic, propionic, and sorbic acids, are used against microorganisms in products with a low pH. Nitrates and nitrites are used to inhibit the bacterium Clostridium botulinum in cured meat products (e.g., ham and bacon). Sulfur dioxide and sulfites are used to control the growth of spoilage microorganisms in dried fruits, fruit juices, and wines. Nisin and natamycin are preservatives produced by microorganisms. Nisin inhibits the growth of some bacteria, while natamycin is active against molds and yeasts.

Foaming and Anti-Foaming Agents

Foaming and Anti-foaming agents are important part of food industry not only for manufacturing of products but also for processing of various food products.

Foaming Agents

Foaming agent also known as aerating agent. A foaming agent is additives that facilitates formation of foam and make it possible to form or maintain a uniform dispersion of a gaseous phase in liquid or solid food. These compounds have been suggested for use in the preparation of certain toppings, cakes and cake mixes, coffee whiteners, whipped creams, beverages and soft drinks.

Propylene Glycol Esters of Fatty Acids

Excellent aerating and foam-stabilising properties in whipped dessert and topping powders. Effective aerating agent in baked goods, especially in combination with distilled monoglycerides and in shortenings. Component and co-emulsifier with distilled monoglycerides in cake improver gels. Widely used in Cake batters, Margarine, Shortening, Powdered desserts, Toppings, Synthetic cream etc.

Glycyrrhizin

Glycyrrhizin is a foaming agent also acts as sweetener and nutrient. It is widely used in baked goods, dairy products (frozen dairy desserts), fruit, vegetables and nut products (hydrolysed vegetable protein (HVP), bean paste, beverages (non-alcoholic beverages), soft drinks (root beer), Sugars, sugar preserves and confectionery (confectionery manufacture, soft candy, confection, frosting, hard candy), alcoholic drinks (liqueurs), vinegar, pickles and sauces (soy sauce) and other (i.e. tobacco, chocolate, vanilla, medicines, gelatin, pudding, chewing gum).

Alternative forms include Calcium glycyrrhizinate, Disodium glycyrrhizinate, Monoammonium glycyrrhizinate, Potassium glycyrrhizinate, Magnesium glycyrrhizinate, Ammonium glycyrrhizinate pentahydrate.

Ammoniated Glycyrrhizin

Widely used in Baked goods, Beverages (alcoholic), Beverages (non-alcoholic), Candy (hard), Candy (soft), Chewing gum, Herbs/ Plant protein products, Seasonings, Vitamin or mineral dietary supplements, Pharmaceuticals.

Function in foods: Flavour enhancer; flavouring agent; surface-active agent; foaming agent; aromatisation of food; sweetening agent for taste correction of food and drugs and production of confectioneries; sweetness potentiators; masking agents; reduce metallic aftertaste from high-intensity sweeteners.

Alginic Acid

Alginates are the principal structural polysaccharide component of brown seaweeds. Alginic acid swells in water, but does not dissolve, and its main applications are in pharmaceutical tablets. In the food industry, it is rarely added directly to food compositions. However, it is produced in situ when sodium alginate is used in acidic food stuffs. In such situations it will form a gel, skin or fibre as a result of its insolubility in water. Alginic acid is used in some formulated alginate products for stabilising ice cream and whipped dairy cream.

Polysorbates

- Polyoxyethylene sorbitan monolaurate (Polysorbate 20).
- Polyoxyethylene sorbitan monopalmitate (Polysorbate 40).
- Polyoxyethylene sorbitan monostearate (Polysorbate 60).

Polysorbate (20, 40 & 60) is widely used within the food industry as a surfactant, for forming oil-in-water emulsions such as dressings, sauces and margarines. The surfactant properties also lead to uses in improving the volume and texture of cakes, the dispersion of coffee whiteners, and the aeration, dryness and texture of whipped cream. It is used to stabilize sauces and dressings and in bakery margarine to improve aeration, cake volume and texture. Typical Products includes cakes and cake mixes, coffee whiteners, whipped creams based on dairy and vegetable fat, margarine, salad dressings and sauces.

Lactic acid Esters of Mono-and Diglycerides of Fatty Acids (Lactems)

Lactems are used in combination with other emulsifiers in whipped topping concentrates, and as aerating and emulsifying agents for cakes and sponges to produce a narrow pore size distribution in the crumb. In mousses these combinations of emulsifiers are used to produce consistent aeration and to maximize volume. Lactems are also used in baking margarines. Typical Products include cakes and whipped toppings.

- Quillaia Extract: Used in Beverage bases; Beverage mixes; Soft drinks.
- Saponin: Saponin is also used in Beverage bases; Beverage mixes; Soft drinks.
- Sodium Bicarbonate: Used in various unstandardized confectionery.
- Others: Apart from above additives, following additives are also known for functioning as foaming agent. Ammonium alginate, Calcium alginate, Calcium stearoyl lactylate, Hydroxypropyl cellulose, Methyl ethyl cellulose, Microcrystalline cellulose (Cellulose gel), Nitrous oxide, Potassium alginate, Propylene glycol alginate, Sodium alginate, Sodium stearoyl lactylate, Xanthan gum.

Anti-Foaming Agents

Also known as defoaming agents. These agents are substances used to reduce foaming caused by proteins or gases that may be interfere with processing. Foaming may be largely suppressed or completely eliminated by the use of small quantities, generally 10 ppm.

These compounds have been suggested for use in the preparation of certain comminuted meat products, bakery products, confections, dairy products, vegetable oils, alcoholic and non-alcoholic beverages, jams and jellies, molasses, soups, starches, syrups and pickles.

Polyoxyethylene Stearate

Polyoxyethylene stearate is made by reacting stearic acid with polyoxyethylene, a polymer of ethylene oxide. Polyoxyethylene stearate is permitted only in wine that has been imported from certain countries, where it is used to inhibit foam formation during fermentation.

Polyoxyethylene Sorbitan Tristearate (Polysorbate 65)

Polysorbate 65 is a tan colour solid produced from a mixture of partial stearate esters of sorbitol and its anhydrides, condensed with ethylene oxide. Polysorbate 65 is used as a surfactant, often in combination with other emulsifiers for forming oil-in-water emulsions. It is used to hold the fat in ice cream to give dry eating characteristics and to retard the development of fat bloom in chocolate products. It is used to reduce foam formation during food processing. Polysorbate 65 has a waxy, somewhat bitter taste. Typical Products include ice cream and frozen desserts, sugar confectionery, cakes and cake mixes, coffee whiteners and whipped creams based on dairy and vegetable fat.

Sodium, Potassium, Magnesium and Calcium Salts of Fatty Acids

The salts of fatty acids are made by reacting the acids with the appropriate hydroxide. The acids used are principally stearic, palmitic and oleic. The salts can be used singly or in mixtures. The salts of fatty acids have a range of uses, usually derived from their fatty acid component. Thus they are defoaming agents, anti-caking and free-flow agents. Magnesium stearate is used to help powders flow during tableting. Other fatty acid salts are used to decrease foam during the processing of beet sugar, chewing gum, and as a yeast activity promoter.

Acetic Acid Esters of Mono-and Diglycerides of Fatty Acids (Acetems)

Acetems are made by reacting mono- and diglycerides of fatty acids with acetic acid in various proportions. In the preparation of jams and marmalades, acetems are used as antifoam to aid filling and present a neat, unbubbled surface. Acetems are also used to manipulate the melting point and plasticity of fats.

Sorbitan Esters

[Sorbitan monostearate, Sorbitan tristearate, Sorbitan monolaurate, Sorbitan monooleate, Sorbitan monopalmitate]. The sorbitan esters are produced by the reaction of the appropriate fatty acid with hexitol anhydride, which is itself derived from sorbitol.

They are used as antifoaming agents in the production of beet sugar, boiled sweets and preserves, and to modify fat crystal structure in chocolate to inhibit the development of the storage defect known as "bloom".

Fatty Acids

The fatty acids include stearic, palmitic and oleic acids. They are made by fractionation of natural fats such as tallow followed by acidification. They can be used alone or in mixtures. Fatty acids have a number of functions, including plasticizers for chewing gum and anti-foaming agents for jams.

Dimethyl Polysiloxane (Silicone, Silicone Oil, Dimethyl Silicone)

Dimethyl polysiloxane is made from silica and oil-derived chemicals. Dimethyl polysiloxane is a surfactant. In foams it occupies the air-water interface and can act in two ways. At low concentration it is a foam stabiliser and at higher concentrations it causes foams to collapse.

Dimethyl polysiloxane is mainly used as an antifoaming agent but is also used to stop hot liquids from sticking to equipment. It is useful in products such as jam, preventing frothing when boiling, and in carbonated drinks, which tend to froth when being filled into bottles or cans.

Dimethyl polysiloxane formulations used as an anti-foaming agent in Apple, Fig, Pineapple marmalade Jam and Jelly; Reconstituted lemon juice; Reconstituted lime juice; Shortening; Skim milk powder and Wine.

Propan-1,2-Diol (Propylene Glycol)

Propane-1,2-diol, also known as propylene glycol, is produced by heating glycerol with sodium hydroxide, or by treating propylene with chlorinated water to form the chlorohydrin, followed by a further treatment with sodium carbonate solution to form the glycol.

It can be used as an anticaking agent, antifoaming agent, emulsifier, flour treatment agent, humectant, stabiliser, thickener, adjuvant and carrier solvent. Polyethylene glycol (molecular weight 3000-9000) is widely used in soft drinks.

Use of Anti-Foaming Agents Permitted in Food Products in India (FSSAI)

Food Products	Dimethyl polysiloxane	Mono and diglycerides of fatty acids and edible oils
Jam / Jellies / Fruit Cheese	10 ppm max	GMP
Fruit Marmalades	10 ppm max	GMP
Culinary Paste / Other Sauces	10 ppm max	10 ppm max
Concentrated Fruit / Veg. Juice / Pulp / Puree	10 ppm max	10 ppm max
Chutney Fruits and / or vegetable / Mango Chutney	10 ppm max	10 ppm max
Green Chilli Paste, Ginger Paste, Garlic Paste, Onion Paste, Whole Chilli Paste	GMP	GMP
Tallow	10 ppm max	-
Lard Edible	10 ppm max	-
Vegetable oils and fats	10 ppm max	-
Thermally Processed Fruits - Pineapple	10 ppm max	

Toxicological Aspect

Very limited information is available on toxicological aspects of foaming and antifoaming agents. Toxicological effects of individual ingredients are discussed below. For foaming Agent i.e. Lactic acid esters, if lactic acid is derived from milk products or by using lactic acid starter cultures that are obtained from milk products, there may be a slight risk for individuals who suffer from milk allergies or lactose intolerance. Patients with a sodium-restricted diet shouldn't abuse sodium bicarbonate. Sodium bicarbonate may cause edema and alkalosis. It can also cause congestive heart failure, hypernatremia, sodium levels higher the normal in blood, and hypertension. In patients with a high calcium diet, the intake of high amounts of sodium bicarbonate may lead to a milk-alkali syndrome, metastatic calcification, kidney stones and renal failure.

For anti-foaming i.e. Dimethyl polysiloxane is non-toxic but will irritate the skin and eyes along with being a danger if inhaled or swallowed in its raw form. Further studies are needed to explore the potential dangers and side-effects of Dimethyl polysiloxane. Propylene glycol in large doses can lead to liver, heart and nervous system disorders as it can damage human cells. Propylene glycol can even disrupt the body's chemical balance, leading to excessive acid production that can cause metabolic problems. After prolonged contact with skin or eyes, it can cause irritation. In high doses, polysorbate 65 can cause cancer if it is contaminated with 1,4- dioaxne, which is readily absorbed by the skin. It is also known to decrease fertility. Polysorbate 65 is also known to be toxic to human organs.

Alkaline Compounds as Food Additive

Alkaline compounds are compounds that raise the pH. Alkaline compounds, such as sodium hydroxide or potassium hydroxide, may be used to neutralize excess acid that can develop in natural or cultured fermented foods. Thus, the acid in cream may partially neutralized prior to churning in the manufacture of butter. If this were not done, the excess acid would result on the development of undesirable flavors.

Sodium carbonate and sodium bicarbonate are used to refine rendered fats. Alkaline compounds are also added to chlorinated drinking water to adjust the pH to high enough levels to control the corrosive effects of chlorine on pipes, equipment, and so forth. Sodium carbonate is also used in conjunction with other compounds to reduce the amount of hardness in drinking water.

Sodium hydroxide is used to modify starch and in the production of caramel. Sodium bicarbonate is used as an ingredient of baking powder, which is used for baked products. It is also a common household item used in a variety of cooking recipe. Alkaline compounds are used in the production of chocolate and in the adjusting of acidity levels in grape juice and other fruit juices that are to be fermented in the production of wine.

Some alkaline compounds, such as sodium bicarbonate are relatively mild and safe to use, while others, such as sodium hydroxide and potassium hydroxide, are relatively powerful reagents and should, not be handled by inexperienced people.

Sensory Agents

Colorants

Colour is an extremely important sensory characteristic of foods; it directly influences the perception of both the flavour and quality of a product. The processing of food can cause degradation or loss of natural pigments in the raw materials. In addition, some formulated products, such as soft drinks, confections, ice cream, and snack foods, require the addition of colouring agents. Colorants are often necessary to produce a uniform product from raw materials that vary in colour intensity. Colorants used as food additives are classified as natural or synthetic. Natural colorants are derived from plant, animal, and mineral sources, while synthetic colorants are primarily petroleum-based chemical compounds.

Soft drink: Many soft drinks, including colas, contain colorants.

Natural Colorants

Most natural colorants are extracts derived from plant tissues. The use of these extracts in the food industry has certain problems associated with it, including the lack of consistent colour intensities, instability upon exposure to light and heat, variability of supply, reactivity with other food components, and addition of secondary flavours and odours. In addition, many are insoluble in water and therefore must be added with an emulsifier in order to achieve an even distribution throughout the food product.

Natural food colorants				
chemical class	Colour	Plant source	Pigment	Pigment
Anthocyanins	Red	Strawberry (Fragaria species)	Pelargonidin 3-glucoside*	Beverages, confections, preserves, fruit products
	Blue	Grape (Vitis species)	Malvidin 3-glucoside*	Beverages
Betacyanins	Red	Beetroot (Beta vulgaris)	Betanin	Dairy products, desserts, icings
Carotenoids**	Yellow/orange	Annatto (Bixa orellana)	Bixin	Dairy products, margarine
	Yellow	Saffron (Crocus sativus)	Crocin	Rice dishes, bakery products
	Red/orange	Paprika (Capsicum annuum)	Capsanthin	Soups, sauces
	Orange	Carrot (Daucus carota)	Beta-carotene	Bakery products, confections
	Red	Mushroom (Cantharellus cinnabarinus)	Canthaxanthin	Sauces, soups, dressings
Phenolics	Orange/yellow	Turmeric (Cuycuma longa)	Curcumin	Dairy products, confections

Synthetic Colorants

Synthetic colorants are water-soluble and are available commercially as powders, pastes, granules, or solutions. Special preparations called lakes are formulated by treating the colorants with aluminum hydroxide. They contain approximately 10 to 40 percent of the synthetic dye and are insoluble in water and organic solvents. Lakes are ideal for use in dry and oil-based products. The stability of synthetic colorants is affected by light, heat, pH, and reducing agents. A number of dyes have been chemically synthesized and approved for usage in various countries. These colorants are designated according to special numbering systems specific to individual countries. For example, the United States uses FD&C numbers (chemicals approved for use in foods, drugs, and cosmetics), and the European Union (EU) uses E numbers.

Common name	United States	European Union	Products
Allura red AC	FD&C red no. 40	...	Gelatin, puddings, dairy products, confections, beverages.
Brilliant blue FCF	FD&C blue no. 1	E133	Beverages, confections, icings, syrups, dairy products
Erythrosine	FD&C red no. 3	E127	Maraschino cherries
Fast green FCF	FD&C green no. 3	...	Beverages, puddings, ice cream, sherbet, confections
Indigo carmine	FD&C blue no. 2	E132	Confections, ice cream, bakery products
Sunset yellow FCF	FD&C yellow no. 6	E110	Bakery products, ice cream, sauces, cereals, beverages
Tartrazine	FD&C yellow no. 5	E102	Beverages, cereals, bakery products, ice cream, sauces

All synthetic colorants have undergone extensive toxicological analysis. Brilliant Blue FCF, Indigo Carmine, Fast Green FCF, and Erythrosine are poorly absorbed and show little toxicity. Extremely high concentrations (greater than 10 percent) of Allura Red AC cause psychotoxicity, and Tartrazine induces hypersensitive reactions in some persons. Synthetic colorants are not universally approved in all countries.

Flavourings

The flavour of food results from the stimulation of the chemical senses of taste and smell by specific food molecules. Taste reception is carried out in specialized cells located in the taste buds. The five basic taste sensations sweet, salty, bitter, sour, and umami are detected in regions of the tongue, mouth, and throat. Taste cells are specific for certain flavour molecules (e.g., sweeteners).

In addition to the basic tastes, the flavouring molecules in food stimulate specific olfactory (smell) cells in the nasal cavity. These cells can detect more than 10,000 different stimuli, thus fine-tuning the flavour sensation of a food.

A flavour additive is a single chemical or blend of chemicals of natural or synthetic origin that provides all or part of the flavour impact of a particular food. These chemicals are added in order to replace flavour lost in processing and to develop new products. Flavourings are the largest group of food additives, with more than 1,200 compounds available for commercial use. Natural flavourings are derived or extracted from plants, spices, herbs, animals, or microbial fermentations. Artificial flavourings are mixtures of synthetic compounds that may be chemically identical to natural flavourings. Artificial flavourings are often used in food products because of the high cost, lack of availability, or insufficient potency of natural flavourings.

Flavour enhancers are compounds that are added to a food in order to supplement or enhance its own natural flavour. The concept of flavour enhancement originated in Asia, where cooks added seaweed to soup stocks in order to provide a richer flavour to certain foods. The flavour-enhancing component of seaweed was identified as the amino acid L-glutamate, and monosodium glutamate (MSG) became the first flavour enhancer to be used commercially. The rich flavour associated with L-glutamate was called umami.

Other compounds that are used as flavour enhancers include the 5′-ribonucleotides, inosine monophosphate (IMP), guanosine monophosphate (GMP), yeast extract, and hydrolyzed vegetable protein. Flavour enhancers may be used in soups, broths, sauces, gravies, flavouring and spice blends, canned and frozen vegetables, and meats.

Sweeteners

Sucrose, or table sugar, is the standard on which the relative sweetness of all other sweeteners is based. Because sucrose provides energy in the form of carbohydrates, it is considered a nutritive sweetener. Other nutritive sweeteners include glucose, fructose, corn syrup, high-fructose corn syrup, and sugar alcohols (e.g., sorbitol, mannitol, and xylitol).

Efforts to chemically synthesize sweeteners began in the late 1800s with the discovery of saccharin. Since then, a number of synthetic compounds have been developed that provide few or no calories or nutrients in the diet and are called nonnutritive sweeteners. These sweeteners have significantly greater sweetening power than sucrose, and therefore a relatively low concentration may be used in food products. In addition to saccharin, the most commonly used nonnutritive sweeteners are cyclamates, aspartame, and acesulfame K.

The sensation of sweetness is transmitted through specific protein molecules, called receptors, located on the surface of specialized taste cells. All sweeteners function by binding to these receptors on the outside of the cells. The increased sweetness of the nonnutritive sweeteners relative to sucrose may be due to either tighter or longer binding of these synthetic compounds to the receptors.

Non-nutritive sweeteners are primarily used for the production of low-calorie products including baked goods, confectioneries, dairy products, desserts, preserves, soft drinks, and table top sweeteners. They are also used as a carbohydrate replacement for persons with diabetes mellitus and in chewing gum and candies to minimize the risk of dental caries (i.e., tooth decay). Unlike nutritive sweeteners, non-nutritive sweeteners do not provide viscosity or texture to products, so bulking agents such as polydextrose are often required for manufacture.

Considerations Required in Use of Food Additives

Before the use of any proposed additives following aspects are required to be taken care.

General Principles for the Use of Additives

- It must be ascertained that the real need exists for the use.
- It does not cause any adverse physiological and harmful effects even upon regular consumption for a long period – must be safe – harmless.
    - It is advisable to take into account Average Daily Intake or equivalent assessment established for the food additive.

- It should not reduce the nutritive value of food.
- It should confirm the agreed specifications.
- Where possible legislation should define permissible maximum quantities of a given additive.
- Even when an additive is considered harmless, it seems advisable (desirable) to use as little as possible of its consistent effect.
- All food additives should be kept under continuous observation and should be re-evaluated wherever necessary, in the light of changing conditions of use and new scientific information.
- Addition of nutrients to food should not be left to the free initiative of the industry, but should in general be prohibited and only in special cases should a dispensation of prohibition be given.

Ascertaining Limits for Addition of Food Additives

Any food additive should be used at the minimum level necessary to produce desired effect. The limit of addition should be established on the basis of:

- The estimated level of consumption of the food for which the additive proposed.
- Minimum levels which in animal studies produce deviations from the normal physiological behavior.
- An adequate margin of safety to reduce risk to a minimum for any hazard to health in all groups of consumers.

Regulations for use and Labeling of Food Additives

Regulations

The use of food additives originated in ancient times but did not engender controversy until the early 1800s, when intentional food adulteration became appallingly common in some countries. Problems with intentional food adulteration continued until about 1920, when regulatory pressures and effective methods of food analysis reduced the frequency and seriousness of food adulteration to acceptable levels in the United States. Since 1920 the use of legally sanctioned food additives has become common.

Regulatory bodies like CODEX Alimentarius Standards (International), Food Safety Standards Act (India) regulates addition of Food additives in Foods by limiting maximum permissible quantity of the ingredient or for some ingredients the limit is not given and it is stated to be added as per GMP as for e.g. use of Lecithin in Sweetened Condensed milk is as per GMP according to CODEX standards.

"All food additives subject to the provisions of this Standard shall be used under conditions of good manufacturing practice, which include the following:

- The quantity of the additive added to food shall be limited to the lowest possible level necessary to accomplish its desired effect.

- The quantity of the additive that becomes a component of food as a result of its use in the manufacturing, processing or packaging of a food and which is not intended to accomplish any physical, or other technical effect in the food itself, is reduced to the extent reasonably possible.
- The additive is prepared and handled in the same way as a food ingredient."

Labeling

According to CODEX general standards Label includes any tag, brand, mark, pictorial or other descriptive matter, written, painted, stencilled, marked, embossed or impressed on, or attached to, a container. Labeling includes the label and any written, printed or graphic matter relating to, and accompanying the food additives. The term does not include bills, invoices and similar material which may accompany the food additives.

According to Food Safety Standards Act, labeling is mandatory for food additives falling in the respective classes and appearing in lists of food additives permitted for use in foods generally, their respective class titles shall be used together with the specific names or recognized international numerical identifications. For externally added flavours, flavouring substances or colour, declaration must be made on the package as it is containing these additives specifying the type either natural, nature identical or synthetic.

Advantages and Disadvantages of Food Additives

There are many benefits of adding additives to the food we eat. It can reduce food spoilage, by keeping the food fresh thus, reducing food loss. It also prevents the growth of bacteria and fungi that cause food poisoning. Food additives can help maintain and improve the nutritional value of food that is sometimes lost during cooking and increase food varieties and convenience as well as improve sensory properties of food. Food additives can meet other special needs required. It will help assure availability of wholesome, appetizing, and affordable food that meet consumers' demand. Food additive promote the development of food industry greatly, and is regarded as the soul of modern or pillar, this basically is to bring many advantages in food industry. Law also controls the use of food additives. An extensive check on the safety of all additives must be carried out before they are allowed in food.

As many benefits as food additives have, however, they also have a good number of disadvantages. Food additives sometimes destroy vitamins in food, replace real ingredients, and may be used to make bad quality food look good. Many fruit drinks make use of food colorings, as do some precooked meats, soy sauce and even beer. The bright colors might make you believe you're eating something healthier, while in reality you might be drinking diluted juice that's full of sugar. Some food additives are toxic if a large amount is used, e.g. sodium benzoate and benzonic acid. Not only that, but, some food additives may be carcinogenic, e.g. sodium nitrate has been shown to cause cancer to mice and may also cause cancer to human beings. Some people are allergic to particular food additives and additives may cause side effects such as allergies, gastric irritation, diarrhea, rashes, asthma, nausea, respiratory irritation and hyperactivity. However, not all people taking these food additives will have these side effects. These side effects differ from person to person and will also vary under different circumstances.

Function of Food Additives

Hundreds of chemical additives are incorporated into foods directly or migrate into foods from the environment or packaging materials. A food additive can be sometimes defined as a substance whose intended use will lead to its incorporation into the food or will affect the characteristics of the foods.

These additives generally provide some types of benefits to the food producer, processor or consumer. For the consumer, additives can improve organoleptic qualities of foods, improve the nutritive value, or ease the preparation of ingredients and meals. Typical additive benefits to the food producer or processor include improving product quality, safety and variety.

Additive may found in varying quantities in foods, perform different functions in foods and ingredients, and function synergistically with other additives. Their functions can usually be classified as one of the following:

- To maintaining or improve nutritional quality.
- To maintain or improve product safety or quality.
- To aid in processing or preparations.
- To enhance sensory characteristics.

Additives that affect nutritional quality are primarily vitamins and minerals. In some foods, these may be added to enrich the food or replace nutrients that may have been lost during processing. In other foods, vitamins and minerals may be added for fortification in order to supplement nutrients that may often be lacking in human diets.

Preservatives or antimicrobial substances are used to prevent bacterial and fungal growth in foods. These additives can delay spoilage or extend the shelf life of the finished product. Antioxidants are additives that also can extend the shelf life of foods by delaying rancidity or lipid oxidation.

Additives that maintain product quality may also ensure food product safety for the consumer. For, example, acids that may be added to prevent the growth of microorganisms that cause spoilage may also prevent the growth of microorganisms that can cause foodborne illness.

Additives that are used as processing or preparation aids, usually affect the texture of ingredients and finished foods. Some of these are classified as emulsifiers, stabilizers, thickeners, leavening agents, humectants and anti-caking agents. Chemical in this group of food additives are also used to adjust the homogeneity, stability and volume of foods.

The fourth major, function of food additives is to enhance the flavor or color of foods to make them more appealing to the consumer. Flavoring chemicals may be used to magnify the original taste or aroma of food ingredients or to restore flavors lost during processing. Natural and artificial coloring substances are added to increase the visual appeal of foods, to distinguish flavors or foods, to increase the intensity of naturally occurring color or to restore color lost during processing.

Safety Aspects and Allergenicity of Food Additives

The contaminants from manufacturing, storing and packaging processes are also considered as indirect food additives. Food additives are generally intended to provide important benefits to the producer or consumer of foods. But sometimes these additives cause unwanted or un-healthful effects. Often these undesirable effects of food additives are due to their excessive or accidental use; usage at an inappropriate stage of production, processing, or storage; or from a lack of purity or quality. Extensive research has been conducted on many additives to show they are safe for consumption and are effective for their prescribed function. For other additives, widespread and long-term use by food processors has demonstrated safety and efficacy.

Whether a chemical is synthetic or "natural" has no bearing on its toxicity. Therefore, this area of food safety concerning naturally occurring contaminants needs to be explored further to determine if the risk from natural contaminants is really low. Naturally occurring seafood toxins and mushroom toxins are relatively common causes of acute food borne disease. The effects of human exposure to natural toxins are difficult to study because consumption of naturally occurring toxins is variable and often cannot be determined. Also, excessive natural toxin consumption generally results in long-term or chronic illness whose source can be difficult to trace.

Both international organizations and local governments generally evaluate the safety of food additives. The goal of local assessment is to take into account local food supply and cultural differences in dietary habits that may influence the intake of food additives.

Food additive intake assessment has three major goals:

- Monitoring the intake of chemicals and relating it to the acceptable daily intake (ADI) values.
- Identifying consumer groups that may be at risk for food additive intake close to or higher than the ADI values.
- Provide information for the regulatory bodies for reassessing the food additive regulations in case of high intake in all or some consumer groups.

Food Allergenicity

Many people have uncomfortable reactions to various foods, and this is often considered to be due to food additives and other artificial chemicals, such as pesticide residues, rather than to the food itself. This discomfort is usually called allergy by nonmedical persons, but there are also nurses and doctors who use the term allergy when they mean various kinds of untoward reactions to foods without knowing the mechanisms of such reactions. Strictly, an allergy is a harmful physiological reaction caused by an immunologic mechanism. If the mechanism is not an allergic one but the reaction resembles allergic reaction, the term intolerance is often used. The symptoms mimic those seen in allergic reactions, but the amount of agent producing the reaction is small enough not to cause a toxic reaction. When the mechanism is not known, it is better to talk about hypersensitivity, which can mean both allergic reaction and intolerance. There are several foods and food additives that can cause both immunologic and nonimmunologic reactions indistinguishable from each other. For example, fish can act both as nonspecific histamine liberators and as true allergens. Tuna, mackerel, and certain cheeses contain histidine and tyrosine to such an extent that histamine and tyramine produced by decarboxylation from them can cause allergic-type symptoms in atopic people. Among food additives, cinnamon and nitrogen mustard are the most well-known causes of both allergic and non-allergic reactions.

Table: Terminology of Hypersensitivity Reactions.

Terminology	Description
Hypersensitivity	A small amount of a substance produces symptoms that can be objectively verified and repeated.
Allergy	Immunologic mechanisms are involved in the pathogenesis of symptoms.
Atopic allergy	The reaction is mediated by immunoglobulin E.
Intolerance	A small amount of a substance produces a reaction similar to or closely resembling a true allergic reaction, but immunologic mechanisms are not involved.

Intolerance and other Nonallergic Reactions

Nonspecific histamine liberation—Cocoa, citrus fruits, strawberry, etc.

- Intestinal diseases—Gluten intolerance produces abdominal and skin symptoms.
- Primary or acquired enzyme deficiency—Diarrhea and colic in lactase deficiency.
- Microbes and their toxins—Bacteria, viruses, yeasts, molds, and fungi cause a diversity of
- Gastrointestinal symptoms.
- Psychological causes may underlie various kinds of skin, gastrointestinal, and other reactions.

Excessive Food Additive Intakes and Population Risk Groups

Currently there is a public debate concerning food additive usage, including food safety, the health risks associated with food additives, and the wholesomeness of food. The food industries assessments suggest that the food currently available is safe if placed in a historical context. The scientific debate regarding additive-associated risks is concerned with the difficulties of epidemiological studies performed in this area, comparisons of the toxicities of processed and unprocessed foods, appraisals of the resource priorities for scientific investigations, and critical evaluations of the methods of analysis. Despite the factors employed to ensure a minimum human sensitivity to the additives in food, particular sections of the population are susceptible to additive associated clinical disorders or toxic responses, some of which are usually thought to have an allergic origin. Excessive intakes have often been suggested for infants and children. The applicability of safety studies has been specifically assessed for infants and children and the different dietary patterns in relation to age and the consequences for intake of food chemical reported.

However, it is likely that the exposure to food additives and the individual's susceptibility expressed as an undesirable response are not separate phenomena, and that the latter could be mediated by allergic reactions. The following considerations should be taken into account when estimating the potential or maximum additive intakes or identifying population subgroups exposed to unacceptable risk:

- When a food additive is known to be present in a food but cannot be detected in analysis, the maximum additive concentration is assumed to be equal to the lower limit of detection. This method enables estimation of the potential average intake.
- Intake estimation can also be achieved by multiplying the highest additive concentration determined by analysis with the amount of the food item or food group under consideration. This value may underestimate the additive intake because consumption of food varies. When the two extreme figures—maximum food additive concentration and maximum food consumption—are combined to give an estimate of intake, the calculated value is probably an overestimate.
- In order to increase the accuracy of estimating the maximum intake value it is necessary to combine the additive concentration of analyzed meals with the true food consumption. Thus, once the population subgroup at risk of excessive food additive intakes is identified, the maximum intakes can be accurately estimated by conducting a duplicate meal study.

Because young men have the greatest demand for food, it may by thought that they also have the highest food additive intakes. However, a comparison of food consumption on a weight-to-body-weight basis indicates that children have the greatest intake of food. Moreover, children consume large quantities of certain food items such as sweets and soft drinks that contain high concentrations of additives. Generally, when food items are identified as being consumed in large quantities by children, the respective additive regulations should be adjusted to permit lower maximum additive levels than those allowed in other foods.

On rare occasions some individuals can experience adverse reactions to food additives. A small percentage of asthmatics can react to sulfites, substances used to prevent certain foods from browning. Also, a very small number of individuals, one or two of every ten thousand, are sensitive to FD&C Yellow No. 5, used as a food coloring, causing itching and hives.

Monosodium glutamate (MSG), commonly found in Chinese foods, can also cause adverse reactions in small groups of people. The symptoms, usually mild, include body tingling or warmth, and chest pain. These symptoms are usually mild and often last less than an hour.

Lastly, people with a rare genetic disease known as phenylketouria (PKU) should avoid foods sweetened with aspartame. Aspartame is made from two amino acids, one being phenylalanine. Individuals with PKU cannot metabolize this amino acid, and if consumed can cause serious side effects including tissue damage. The best advice to any individual that has adverse reactions to any food additives is to read labels carefully and avoid these products whenever possible. If an adverse reaction does occur, be sure to contact your physician immediately.

Regulation of Food Additive Usage

Safety of Food Additives

Food additives approved by the FDA are considered to be safe for human consumption, and many processed food manufacturers claim that there is no solid evidence to show direct association between food additives and human health. However, the U.S. government has claimed that safety aspects of food additives are not fully known. "Any substance the intended use of which results or may reasonably be expected to result, directly or indirectly, in its becoming a component or otherwise affecting the characteristics of any food. Such substance is not generally recognized, among experts qualified by scientific training and experience to evaluate its safety, as having been adequately shown through scientific procedures. to be safe under the conditions of its intended use". Thus, with uncertainty about the safety of the consumption, the safety of the food additives and preservatives should be carefully examined once more before eating them.

In addition to the previously mentioned health problems of consuming modern synthetic preservatives, other food and color additives have been linked with allergic reactions, cancer, asthma, and birth defects. For example, sulfites used to prevent discoloration are shown to cause allergic reactions, according to the FDA. Once a person develops sulfite allergies, it can potentially lead to fatal respiratory distress.

BHA and BHT is most frequently used food additive in processed foods e.g. in processed cereals. These chemicals are usually added to prevent oxidation of fats and oils in food. Oxygen tends to react with BHA and BHT before oxidizing fats, which in turn keeps the food from going rancid. Although the United States Food and Drug Administration (FDA) have approved BHA and BHT along with about 3,000 food additives for consumption, they have been shown to cause a number of health problems. Some studies claim that synthetic preservatives worsen Attention Deficit Disorder (ADD) and Attention Deficit Hyperactivity Disorder (ADHD) symptoms in those affected. Other studies also show that certain persons may have more difficulty with digesting and metabolizing the compounds of BHA and BHT, which results in behavioral changes and other health problems.

Additionally, most fruit juices that are often marketed to parents of young children contain additives, including preservatives, artificial sweeteners and colorings. A study reports that preservatives such as sodium benzoates may cause increased hyperactivity in 3-year-old and 8/9-year-old children. The increase of the hyperactivity was about 50% greater for those children who regularly drank fruit juices with additives than those who drank juice without additives.

Concern about the possible effects of food additives on health, including cancer, is one reason that many people are now interested in organic foods. Organic foods are often promoted as an alternative to foods grown with conventional methods that use chemical pesticides and herbicides, hormones, or antibiotics. These compounds cannot be used for foods labeled as "organic." Organic foods, as defined by the US Department of Agriculture (USDA), also exclude genetically modified foods or foods that have been irradiated.

Food Safety Regulations

Today, food and color additives are more strictly studied, regulated and monitored more than any other time in history. The FDA sets safety standards, determining whether a substance is safe for its intended use. Rigorous testing is done to determine the amount of safe levels of different food additives. Additionally, food manufacturers must prove to the FDA their product is safe before it is put on the market.

The majority of direct food ingredients are used on the basis of a determination that they are GRAS or prior sanctioned. Review of some items on the GRAS list has indicated that the majority present no significant hazard with normal use, although only a small percentage have been thoroughly evaluated. The other direct food additives used in foods have been approved, and their uses are regulated by the FDA.

Food safety in India is ensured by Government of India's Ministry of Health under the provisions of Prevention of Food Adulteration Act & Rules previously, which at present is under Food Safety Standards Act. They are responsible for Food Laws and the rules therein. State government, Food & Drug Administration (FDA), which carries out surveillance using food inspectors, does the enforcement. There are food analysis laboratories, both state and central, which verify the authenticity of food products. Any food safety legislation or standard, requires involvement of several aspects including Research & Development, Information & Documentation, Education & Training, Quality Assurance Program, Codex & International Norms, Advisory System, Planning, Enforcement and Surveillance. Various activities take place at different places such as education & research institutions, government laboratories, data bases including international & national, industry production and quality evaluation centers, and finally state level enforcement and surveillance departments. Due to the complex nature, any change in standards and enforcement has to be properly planned and executed after careful consideration of all these factors.

There are different sets of regulations everywhere. Each country has its own set of rules for regulating food additives for example, US FDA Guidelines & Regulations gives the American regulations for food additives. Thus anyone producing and marketing food products in the US must abide by them. India has its own set of regulations under Food Safety Standards Act. Each country has a set of regulations. When an Indian company wants to export to US, then it will have to follow the US regulations. When it wants to export to Australia their rules have to be followed. So there might be difficulties trying to follow many sets of regulations.

A group of countries may have a common regulation for example, European Union Directives, which give regulations for countries affiliated to it. This allows free exchange of food products across those EU countries. It avoids confusion because of many different regulations being followed for different countries. For international trade we have Codex, SPS, TBT regulations. Under the WTO agreements, common regulations have been arrived at for those countries signatories to the agreement and this allows the international trade without much problems. FAO/WHO has come up with Codex rules, which are accepted by these countries.

Food Additives: Approval Process

Any new additive before approving must undergo rigorous toxicity studies, including acute and chronic studies involving biochemical evaluation, teratogenic studies, reproductive studies besides the LD50 tests. In the US, Delaney Clause governs the approval of any food additive, under which the additive is banned if found to be carcinogenic, under any condition or level, a very difficult zero risk condition.

However, exposure assessment is very important in determining the risk involving any additive under the modern practice of determining safety. The Risk Analysis, adopted nowadays involves, risk assessment, wherein the Hazard is identified & characterized, Exposure is assessed and thus risk is characterized. Once the Risk is assessed, it must then be managed so hazardous conditions do not arise. Finally the risk must then be communicated.

General Principles of Food Safety Risk Management

- Risk management should follow a structured approach.
- Protection of human health should be the primary consideration in risk management decisions.
- Risk management decisions and practices should be transparent.
- Determination of risk assessment policy should be included as a specific component of risk management.
- Risk management should ensure the scientific integrity of the risk assessment process by maintaining the functional separation on risk management and risk assessment.
- Risk management decisions should include clear, interactive communication with customers and other interested parties in all aspects of the process.
- Risk management should be a continuing process that takes into account all newly generated data in the evaluation and review of risk management decisions.

Regulatory Authority

Numerous research studies have confirmed that appropriate use of many additives is safe for human health and provides a benefit to the processor, preparer, or consumer. Other additives are considered safe and efficacious due to successful widespread use over many years. Multiple federal agencies, laws, and regulations work together to ensure the safety and efficacy of thousands of food additives, but there are great differences in how many of these additives are regulated. The Food, Drug and Cosmetic Act of 1938 gives the U.S. Food and Drug Administration authority to regulate foods, ingredients, and their labeling.

The 1958 Food Additives Amendment to the FD&C Act requires FDA approval for the use of new additives prior to its inclusion in food. Also this Amendment requires the additive manufacturer to prove an additive's safety for its recommended use. Food additives are defined as substances which may, by their intended uses, become components of food, either directly or indirectly, or which may otherwise affect the characteristics of the food. The term specifically includes any substance intended for use in producing, manufacturing, packing, processing, preparing, treating, packaging, transporting, or holding the food, and any source of radiation intended for any such use. This definition does not include some classes of additives such as pesticide chemicals for raw agricultural products, new animal drugs, certified colors, or colors exempt from certification. These additives are similarly regulated, but under other laws or acts. Additionally, food additives that have a history of safe usage are exempted from the regulation process described in the Food Additives Amendment. One group of these additives is known as "generally recognized as safe" (e.g., salt, vitamins). They have been generally recognized, by experts, as safe based on their extensive history of use in foods before 1958, or based on published scientific evidence. Another group of exempted additives is designated as prior sanctioned additives. These are substances that the FDA or U.S. Department of Agriculture (USDA) had determined were safe for use in specific foods prior to 1958 (e.g., sodium nitrite to preserve meat). Other important federal regulation of food and color additives is described in the 1960 Color Additive Amendment to the FD&C Act, the 1990 Nutrition Labeling and Education Act, and the 1996 Food Quality Protection Act. Current good manufacturing practice (CGMP) regulations can limit the quantity of food and color additives used in production. Manufacturers may only use the amount of an additive necessary to achieve a desired effect.

In the case of pesticides for foods and animal feeds, the U.S. Environmental Protection Agency and the FDA have regulatory responsibilities for the approval and monitoring of pesticide use. Three major legislative acts describe the authority of the EPA and FDA to regulate pesticide chemicals. The Federal Insecticide, Fungicide and Ro-denticide Act prescribes that the EPA is responsible for registering or licensing pesticide products for use in the United States. The Federal Food, Drug and Cosmetic Act govern the maximum level of pesticide residues allowed in or on specific human foods and animal feeds. In 1996, the Food Quality Protection Act (FQPA) amended these two acts in several respects. For example, to assess the risks of pesticide residues in foods or feeds, the FQPA requires that the combined exposure from dietary and other non-occupational sources be considered. Also, when setting new or adjusted tolerances for pesticides, the EPA must consider any special risks to infants and children.

Monitoring

Federal regulations are in place to determine if an additive is safe and effective under its prescribed uses. Also, regulations may stipulate the permitted food uses or usage levels for hundreds of additives. Nevertheless, the usage and labelling of food additives may be monitored through analytical testing and label review of finished products or ingredients. Improved chemical analytical methodology has resulted in faster extraction and identification of additives from foods. Many testing protocols can now detect chemical concentrations of very low levels (e.g., parts per billion) and with increasing discrimination.

Labeling

The risks or benefits of food additives and ingredients must be clearly displayed for consumers. The FD&C Act requires, in virtually all cases, a complete listing of all the ingredients of a food. The Nutrition Labeling and Education Act, which amended the FD&C Act, requires most foods to bear nutrition labeling and requires food labels that bear nutrient content claims and certain health messages to comply with specific requirements. Two of the exemptions from ingredient labeling requirements have resulted in special product labeling efforts to protect the health of consumers. First, the act provides that spices, flavorings, and colorings may be declared collectively without naming each one. One exception is the artificial color additive FD&C Yellow No.5. This chemical must be specifically identified in the ingredients statement of finished foods because a small percentage of the population may be allergic or sensitive to the additive. Second, FDA regulations exempt from ingredient declaration incidental additives, such as processing aids, that are present in a food at insignificant levels and that do not have a technical or functional effect in the finished food. One important example of an incidental additive is peanuts. An increasing number of products are identified that they "may contain peanuts." While peanuts or peanut-derived ingredients were not intentionally added to these products, residues from peanut use in processing on nearby equipment or previous production runs may have contaminated these products with peanut residues. Since peanuts are one of the leading causes of allergenic responses to foods, many companies have chosen to label some products with "may contain peanuts." Some processors have elected to process products with peanuts in separate locations from their product lines that are not to contain peanuts.

Some foods may be identified or labeled to contain additives that can improve public or individual health. Many ready-to-eat breakfast cereal products are fortified with several vitamins and minerals. A quantity of these substances is added to the cereal so that a consumer may expect to consume 25 to 100% of the recommended daily intake of that nutrient from a defined size serving. Additives that can improve human health are sometimes advertised elsewhere on finished product packaging besides the ingredients list and nutritional label. Advertising claims that imply better health through consumption of a food product additive are permitted for specific additives, ingredients, or inherent compounds. As one example, regular consumption of folic acid can be advertised as beneficial for preventing neural tube defects in new born babies.

Allergic people who are highly sensitive to soy, rape, egg or sunflower seeds might react to lecithin. The origin of lecithins derived from the most common allergenic foods (as defined in local legislation) must be declared on food labels. Hypersensitivity to sulphites is relatively well described, especially in people with asthma, and they may also trigger skin reactions such as hives (urticaria). When used at levels of or above 10mg per kg, sulphites have to be labelled on pre-packed foods within the EC.

References

- Food-additive-50529: britannica.com, Retrieved 10, June 2020
- Functions-of-food-additives: foodadditive.blogspot.com, Retrieved 03, Feb 2020
- Preservatives, food-additive-50534: britannica.com, Retrieved 02, March 2020
- Advantages-and-disadvantage: foodadditivesbasic.weebly.com, Retrieved 05, May 2020
- Alkaline-compounds-as-food-additive: foodadditive.blogspot.com, Retrieved 17, August 2020

2

Food Coloring and Flavoring

Any pigment added to a food item to enhance its color is referred to as food coloring. The ingredients that are added to any food item to improve its taste or flavor are termed as food flavoring. This chapter has been carefully written to provide an easy understanding of food coloring, classification of colorants and food flavoring.

Food Coloring

Food coloring or colouring is any of numerous dyes, pigments, or other additives used to enhance the ap-pearance of fresh and processed foods. Colouring ingredients include natural colours, derived primarily from vegetable sources and sometimes called vegetable dyes; inorganic pigments; com-binations of organic and metallic compounds (called lakes); and synthetic coaltar substances. They are added to orange and potato skins, sausage casings, baked goods, candies, carbonated drinks, gelatin desserts, powdered drink mixes, and many other foods. Many of these additives are also employed as colouring agents in cosmetics, drugs, and products such as toothpaste and mouthwash.

Importance of Food Colors

1. Color has always had an important implication on the minds of people as far as food is concerned. Cuisines prepared in attractive colors have immensely lured men folk in all the quarters of the world. It is therefore very necessary to preserve the natural or maintain the characteristic color of a food product while it is manufactured or stored for future use. A non-attractive color however makes the food look un-fresh and is likely to be rejected.

2. The color of food is a significant factor in determining its acceptance. We expect to see food looking natural. Many consumers associate natural products with superior quality and a good, natural looking color in a food or beverage will signal high quality while a washed out or artificially bright product can give the opposite impression.

3. Natural appearance is appetizing. We become cautious when a food shows an unexpected color, interpreting it as a possible sign of spoilage or poor processing.

4. Color in food and drink products is important to the manufacturer primarily because it is important to his customers. They live in a world full of color; they eat food which is colored by nature and expect the foods produced by manufacturers to be similarly colored as appropriate. Foods and drinks throughout the world particularly vegetables and fruits abound in color. Chefs roast, toast and bake dishes to give an attractive color. Beverages are processed to give certain characteristic color.

5. The contribution to the appearance and attractiveness of foods is paramount, partly because we expect foods to have their own characteristic appearance including color and partly because color is important in the perception of flavor. We can say that color can label flavor type. In prepared foods deriving from nature we expect the color to be associated with the flavor so that asparagus soup should be green and raspberry ice-cream is expected to be red.

6. Adding colors to food is also important to offset the color loss due to light, air, extreme temperature, moisture and processing of food.

7. Color can also assist in the use or measurement of products. It indicates whether the ingredients have been properly mixed or not like in cake batter.

8. It is also important to provide a colorful appearance to certain "Fun Food" to attract consumers especially children. Candies are colored to create a festive appearance.

9. Color can also give some assistance in how much to use. It can help us in judging the quantity of margarine spread on white bread or of orange squash to water.

Classification of Colorants

Directives on colorants are examined in 3 groups. These are as follows:

1. Colorants whose ADI values are determined and allowed for use.

2. Colorants permitted to use only for special purposes (such as surface finishing) ($CaCO_3$, aluminum, silver, gold).

3. Colors that are only allowed to use in some foods (Titanium dioxide, vegetable carbon, red beet).

The use of colorants outside of this classification is banned. Furthermore, despite the ban on the addition of colorants to products such as mineral waters, milk, flour and tomato paste, different implementations are applied in national legislation. Colorants differ from each other by various properties such as chemical structures, sources, and purpose of use. As it is difficult to classify the colorants according to these properties, they are divided into two groups based on their sources as natural and synthetic.

Natural Colours

The natural pigments may be divided into four, not necessarily homogenous groups:

- Isoprenoid derivatives - Carotenoids.
- Porphyrins - Chlorophylls and hemes.
- Phenolics - Anthocyanins, betalains, quinones, curcuminoids.
- Miscellaneous naturally occurring colorants - Riboflavins, caramels, melanoidins and melanins.

Natural pigments are formulated either as dyes (hydrophilic powders or lipophilic oleoresins) or lakes.

Use of Natural Colours in few Countries

The natural colorants included in the United States (US) category are annatto extract, β-apo-8′-carotenal, β-carotene, beet powder, canthaxanthin, caramel color, carrot oil, cochineal extract (carmine), cottonseed flour (toasted, partially defatted, and cooked), ferrous gluconate, fruit juice, grape color extract, grape skin extract (enocyanin), paprika, paprika oleoresin, riboflavin, saffron, titanium dioxide, turmeric, turmeric oleoresin, vegetable juice. Examples of some exempt natural colourings include grape skin extract for restricted use only in beverages; marigold (Tagetes) meal can be used only for chicken food. The list of natural colourants with their specific properties is shown in Table. The properties of natural colourants is highlighted in Table.

Table: List of natural colourants with their specific properties.

Natural food colourant	EU code	Colour hue	Source and properties
Curcumin, turmeric, turmeric oleoresin	E 100	Orange-yellow	Plant rhizome (*Curcuma longa*); strong odor and bitter, sharp taste
Riboflavin (ii) riboflavin-5-phosphate	E 101	Yellow-green	Semi-synthetic, obtained from bacterial fermentation (*Blakeslea trispora* and *Ashbya gossypi*); fluorescent lightsensitive, bitter taste
Carminic acid and carmine, Cochineal extract	E 120	Orange to red, pink to red	Female cochineal insect
Chlorophyll grass	E 140	Green to olive	Lucerne and nettle; labile to photo-oxidation
Chlorophyllins (copper complex of chlorophyll)	E 141	Bluish green	Grass, lucerne, and nettle; not claimed as natural on food labels
Caramel colours (I, II – sulfite, III – ammonia, IV – ammonia sulfite)	E 150a-d	Brown	Food-grade carbohydrates
Carbo medicinalis (vegetable carbon)	E 153	Black	Burned plant material
Carotenes (mixtures including β-carotene), β-carotene	E 160a	Yellow to orange	Palm oil or fermentation of *Blakeslea trispora* and *Ashbya gossypi*; good pH stability, oil-soluble, easily oxidized
Bixin, Norbixin Annatto extracts	E 160b	Orange-red	Seeds of annatto (*Bixa orellana*)
Capsanthin, capsorubin, paprika, paprika oleoresin	E 160c	Reddish (Carmine) orange	Paprika (*Capsicum annum L.*)
Lycopene	E 160d	Orange red	Tomato (*Lycopersicum esculentum*)
β-Apo-8-carotenal	E 160e	Orangish red	Semi-synthetic, from β-carotene
Ethyl ester of β-apo-8-carotenoic acid	E 160f	Orangish red	Semi-synthetic, from β-carotene
Canthaxanthin	E 161a	Orange and pink	Salmon, shrimp; semi-synthetic from β-carotene

Lutein	E 161b	Orange and yellow	Aztec marigold *(Tagetes sp.)* petals
Betanin, Dehydrated beets	E 162	Pink to red	Red beet root; heat, light and oxy-gen-sensitive
Anthocyanins, grape color extract, grape skin extract	E 163	Pink-red to mauve-blue	Elderberries, black grape skins, black carrots, red cabbages; pH-dependent, heat- and oxidation-sensitive
Calcium carbonate	E 170	White	Mineral rocks; surface dye
Titanium dioxide	E 171	White	Mineral rocks; surface dye
Iron oxides and hydroxides, synthetic	E 172	Yellow, red, black	Mineral rocks; surface dye
Aluminium	E 173	White	Mineral rocks; surface dye
Silver	E 174	White	Surface dye
Gold	E 175	Yellow	Surface dye
Rubin BK	E 180	Red	Surface dye
Paprika	No	Red	Paprika
Saffron	No	Yellow-orange	Saffron

The detailed description of few natural colourants is given below:

Caramel

Caramel color (E 150) is a dark brown or even black product used for centuries in home cooking to provide color and specific aromas to foods. Caramel shall be prepared from the food grade carbohydrates or their combinations (sucrose, glucose, fructose, invert sugar, lactose, malt syrup, molasses, starch hydrolysates and fractions thereof and/or polymer thereof) in the presence of food grade acids, alkalis or salts. According to the Food and Drug Administration (FDA), caramel is the dark-brown liquid or solid material resulting from the carefully controlled heat treatment of the food-grade carbohydrates namely dextrose, invert sugar, lactose, malt syrup, molasses, starch hydrolysates and fractions thereof, sucrose.

- According to FSSAI, the four types of caramel include:
- Type-I: Plain caramel shall be prepared by heating carbohydrates with or without acids or alkalis, or their salts. Ammonium or sulphite compounds are not used.
- Type-II: Caustic sulphite caramel shall be prepared by heating carbohydrates with or without acids or alkalis or their salt in the presence of sulphite compounds; ammonium compounds are not used.
- Type – III: Ammonia process caramel shall be prepared by heating carbohydrates with or without acids or alkalis or their salts in the presence of ammonium compounds; sulphites are not used.
- Type-IV: Ammonia sulphite caramel shall be prepared by heating carbohydrates with or without acids or alkalis or their salts in the presence of both sulphite and ammonium compounds.

Table: Properties of natural colourants used in food industry.

Colourant	Colour	Mol. Wt.	Melting point	Solubility
β-carotene	Yellow	536.9	183 °C	Soluble in carbon disulphide, benzene and chloroform, moderately soluble in normal hexane, cyclohexane, ether, petroleum ether and oils; practically insoluble in methanol; insoluble in water.
Chlorophyll	Green	893.5 (for a) 907.5 (for b)		Soluble in ethanol, ether, chloroform and benzene; insoluble in water.
Caramel	Dealt separately			
Annatto	Natural orange	394.5 (Bixin380.5 (Norbixin)		Norbixin is water soluble, bixin is fat soluble.
Riboflavin	Yellow to Orange-yellow	376.4	280 °C	Slightly soluble in water, more soluble in saline solution and in 10 % (w/v) solution of urea, sparingly soluble in alcohol, practically insoluble in chloroform and ether; but soluble in dilute solution of alkali hydroxides.

Four classes of caramels are recognized depending on the accelerator used: burnt sugar, caustic, ammonia, and ammonium sulfite. The first is used mainly as a flavoring additive, and the other as food colorants.

Caramel colors are the most widely used food coloring agents found in a wide range of foods and beverages. Caramel represents a significant segment of the color market; about 11% of total food color production. It is obtained by heating sugars in various conditions. Caramel is the colorant soluble in water and insoluble in organic solvents.

The US Food and Drug Administration (FDA) permits the use of caramel color in food in general. Over 80% of the caramel produced in the United States is used to color soft drinks, particularly colas and root beers.

Carotenoids

The most commonly used natural carotenoid extracts for foodstuffs are annatto, paprika, and saffron. Many other sources, including alfalfa, carrot, tomato, citrus peel, and palm oil, are also utilized. The properties of carotenoid colourants with their vitamin A activity is shown in table.

- Annatto [E 160(b)] is the orange-yellow, oil-soluble natural pigment extracted from the pericarp of the seed of the Bixa orellana L. tree. The major coloring component of this extract is the diapocarotenoid bixin, Several other pigments, mainly degradation products of bixin, are also present, including transbixin, norbixin, and trans-norbixin. Bixin is a methyl ester of a dibasic fatty acid, which on treatment with alkalis is hydrolyzed to water-soluble norbixin. Two types of annatto are therefore available: the extract in oil containing bixin, and a powdered, water-soluble form containing norbixin.
- Annatto butter color: Annatto extract in oil, as solution or suspension, is prepared by extraction of the outer coating of seeds with vegetable oils (singly or mixtures). The colour of

annatto solution in amyl acetate at a dilution of 1:1000 (m/v), when measured in a lovibond tintometer with a 1 cm cell spectrophotometrically/calorimeterically shall be as follows: Yellow units (Min.) - 5.0; Red units (Min.) - 0.4.

- Annatto cheese colour: Water soluble annatto colour is prepared by extraction of the outer coating of the seeds with aqueous alkali (NaOH or KOH – 0.5-3.0%). The colour of the solution in 0.1 N NaOH or KOH at a dilution of 1:1000 (m/v) measured in a 1-cm cell shall be: Yellow units (Min.) – 5.0; Red units (Min.) – 0.4.
- Paprika oleoresin [E 160(c)]: is the orange-red, oil-soluble extract from sweet red peppers (Capsicum annum). The major coloring compounds are xanthophylls: capsanthin, capsorubin as their dilaurate esters, and β-carotene. The presence of characteristic flavoring and spicy pungency components limits application of this extract in foodstuffs.
- Raw, unrefined palm oil contains 0.05 to 0.2% carotenoids with α- and β-carotenes, in a ratio of 2:3, as the main constituents. It is particularly used as a colorant for margarine.
- Saffron: It is an extract of flowers of Crocus sativus, contains the water-soluble pigment crocin, the digentiobioside of apocarotenic acid (crocetin), as well as zeaxanthin, β-carotene, and characteristic flavoring compounds. The yellow color of this pigment is attractive in beverages, cakes, and other bakery products. However, use of this colorant is restricted by its high price.
- Carrot extracts (E 160a): Carrot extracts, carrot oil, and related plant extracts are also available in the market. Their main components are β- and α-carotenes have been developed. Purified crystalline products contain 20% α-carotene and 80% β-carotene. They may be used for coloring fat-based products, such as dispersion of microcrystals in oil.
- Individual carotenoid compounds such as β-carotene, β-apo-8′-carotenal, β-apo-8’-carotenoic acid ethyl ester, and canthaxanthin are synthesized for use as food colorants for edible fats and oils. The carotenoid pigments, in combination with surface active agents, are also available as micro-emulsions for coloring foods with high water content.

Table: Properties and vitamin A activity of carotenoid colorants.

Carotenoid	Color	Solubility in oil (g/100 cm3) at 20oC	λmax	Vitamin A activity (IU/mg)
β-carotene	Yellow	0.05-0.08	455-456	1.67
β-Apo-8′-carotenic acid ethyl ester	Yellow to Orange	0.70	448-450	1.20
β-Apo-8′-carotenal	Orange to Red	0.70-1.50	460-462	1.20
Canthaxanthin	Red	0.005	468-472	0

Turmeric and Curcumin

Turmeric is an aromatic spice obtained from the dried ground rhizomes of Curcuma longa L., a perennial shrub. Curcuma longa L has bright green leaves, conical yellow flowers, and reaches maturity after 7 to 10 months, when rhizomes are harvested. The dried ground rhizomes yield a bright yellow powder also known as yellow ginger or Indian saffron. It has the special property of imparting both color and flavor.

The main compounds involved in color are curcumin, demethoxycurcumin, and bisdemethoxycurcumin. The three commercial forms based on Curcuma longa L. are turmeric powder, turmeric oleoresin, and curcumin powder. Commercially available products called curcumins contain curcumin (1,7-bis(4-hydroxy-methoxyphenyl)-1,6-heptadiene-3,5-dione) as the major component (~ 77% of total curcuminoids). Turmeric and curcumin are insoluble in water but soluble in alkalis, alcohols, and glacial acetic acid, and are used mainly as food color and secondarily as a spice. All turmeric pigments have good heat stability, but they are light sensitive.

Turmeric is used mainly as a spice, to give specific flavor and color, but also as an additive for maintaining freshness and improving the palatability and shelf lives of perishable foods. Turmeric powder is obtained from dried rhizomes. The turmeric powder standardized with maltodextrin contains 8 to 9% curcumin. The stability of turmeric under light and alkaline conditions have been increased by the addition of acids (gallic, citric, gentisic) as stabilizers. The oleoresin is obtained from turmeric powder by solvent extraction; the solvents used are ethyl acetate, acetone, carbon dioxide, dichloromethane, n-butanol, methanol, ethanol, and hexane and their mixtures; isopropanol and trichloroethylene is also one good mixture. After filtration the solvents must be completely removed from the oleoresin.

Turmeric powder and turmeric oleoresin contain pigments, flavor compounds (volatile oils), resins, and fats. Both can be used as spices and as food colorants. Curcumin powder (E 100) is obtained from the turmeric oleoresin by crystallization. It appears as an orange-yellow crystalline powder with a melting point at 179 to 182 °C. It is soluble in ethanol, propylene glycol, and acetone and insoluble in water. Curcumin powder must contain minimum of 90% pigment.

Generally turmeric powder is used in mustard paste and curry powder as a colorant and for aroma. Oleoresin is added to mayonnaise, to breading of fish and potato croquettes, and to non-alcoholic beverages. However, curcumin (E 100) is added to products where turmeric is incompatible, such as cheese, butter, ice cream, and some beverages.

Other applications are coloring mustard, pickles, mayonnaise, salad dressings, oils, and cauliflower. In the European Union (EU), curcumin is permitted in alcoholic beverages (quantum satis), jam, jellies, marmalades (100 mg/kg), sausages, pâtés (20 mg/kg), dried potato granules and flakes, non-alcoholic drinks, and confectionery. The FDA approved turmeric powder (CI 73600) and turmeric oleoresin (CI 73615) but not curcumin powder for general use in foods. Turmeric powder is used in the 0.2 to 60 ppm range and the oleoresin in the 2 to 640 ppm range. According to Joint Expert Committee on Food Additives (JECFA), the acceptable daily intake (ADI) was established at 0 to 3 mg kg/body weight. Curcumin is a potent candidate to replace the synthetic tartrazine pigment.

Chlorophyll

Chlorophyll, the green pigment of plants, is extracted and widely used as a colouring matter for various food items. In higher plants and algae, except the blue-greens, chlorophyll is found in chloroplasts, while in blue-green algae and photosynthetic bacteria, it is located on intracellular lamellae. In living plant tissues, in chloroplasts, chlorophyll is complexed with polypeptides, phospholipids, and tocopherols and held within hydrophobic membranes.

Typical leaf material contains about 2.5 mg/g total chlorophylls, 0.3 mg/g xanthophylls, and 0.15 mg/g carotenes. In many fruits, chlorophyll is present in the unripe state, and gradually disappears during ripening as the yellow and red carotenoids take over.

Chlorophylls are derivatives of dihydroporphirin chelated with a centrally located Mg ion. Cu or Zn ions retain the green color. The chlorophyll preparations for the food colorant market are mainly obtained from alfalfa (Medicago sativa) and nettles (Urtica dioica). Brown seaweeds are a good source of chlorophyll because as in single-cell phytoplankton, they contain chlorophyll c, which is more stable than chlorophyll a and chlorophyll b. Acetone, methanol, ethanol, and chlorinated solvents are used as extracting vehicles. The yield of extraction is ~20% in which chlorophylls, pheophytins, and other degradation products are included.

Chlorophylls are approved in Codex legislation as food additives for different applications. The oil-soluble chlorophylls, as well as the water-soluble chlorophyllins, a semi-synthetic mixture of Na and Cu salts derived from chlorophyll, have good stability in light and heat, and moderate stability with respect to both acid and alkalis. Their application as food colors is in canned products, confectionery, soups, and dairy products.

Anthocyanins

They are present in almost all higher plants and are the dominant pigments in many fruits and flowers, giving them red, violet, or blue color. Substitution of the hydroxyl and methoxyl groups affects the color of the anthocyanins. An increase in the number of hydroxyl groups tends to deepen the color to a more bluish shade. An increase in the number of methoxyl groups increases redness. In the presence of metals, anthocyanins can form purplish-blue or slate-gray pigments called lakes.

The examples of naturally occurring anthocyanidins include Pelargonidin (Orange color), cyaniding, Peonidin (orange-red color), Delphinidin, Petunidin and Malvidin (bluish-red color). The fruits potent in anthocyanins include Blackberry, Bilberry, Blackcurrant, Chokeberry, Cranberry, Strawberry and Red grapes.

Anthocyanins appear to have low stability in all products manufactured from fruits. This limits the use of these pigments as food colorants. Anthocyanin concentrate may be used as a food colorant at a pH less than 4.

Betalains

These consist of red-violet betacyanins (λmax ~ 540 nm) and yellow betaxanthin (λmax ~ 480 nm). The betanin, glucoside of betanidin, is the main betalain pigment in red beetroot and accounts for 75 to 95% of total pigments in red beets. The other red pigments are isobetanin (C-15 epimer of betanin), prebetanin and isoprebetanin. The latter two are sulfate monoesters of betanin and isobetanin, respectively. The major yellow pigments are vulgaxanthin I and II. High betalain content in beetroot, on an average 1% of the total solids, makes these vegetables a valuable source of food colorants. The color stability of betanin solution is influenced by pH and heating. Betanin is stable at pH 4 to 6, but thermostability is greatest between pH 4 and 5. Light and air have a degrading effect on betanin.

Beetroot red (E162), available as liquid beetroot concentrate and as beetroot concentrate powders, is suitable for products of relatively short shelf life, which do not undergo severe heat treatment, e.g., meat and soybean protein products, ice cream, and gelatin desserts.

Red beet betalains consisting of betanin, isobetanin, prebetanin, and smaller quantities of vulgaxanthin I and vulgaxanthin II, respectively, were proposed as viable alternatives to synthetic colorants, especially in low-acid foods such as dairy and meat products. In the latter commodities, red beet was even considered as a replacement for nitrite to achieve reddening, but it lacks the antimicrobial activity of nitrite. Moreover, red beet addition contributes to higher nitrate levels of the particular food, which represents a potential source of nitrite.

Betalains possess high molar extinction coefficients and their coloring power is comparable to that of synthetic colorants. Compared with the latter, betalains are not toxic, nor do they cause allergic reactions.

Riboflavin

Riboflavin (E 101) is an orange-yellow crystalline powder, intensely bitter tasting, which is very slightly soluble in water and ethanol, affording a bright green-yellow fluorescent solution. Riboflavin is stable under acid conditions, but unstable in alkaline solution and when exposed to light. Reduction produces a colorless leuco form, but color is regenerated again in contact with air.

Riboflavin-5′-phosphate sodium salt is much more soluble in water than unesterified riboflavin, and is not so intensely bitter. It is one of the physiologically active forms of vitamin B_2. It is rather more unstable to light than riboflavin. Both forms can be used as coloring and as an enriching food additive to cereal, dressing, and cheese.

Cochineal, Carmine

Cochineal (E 120 – Quinone pigment) is the red coloring matter extracted from the dried bodies of female insects of the species Dactylopius coccus Costa or Coccus cacti L. These insects are cultivated on cactus plants in Peru, Equador, Guatemala and Mexico. The major pigment of cochineal is polyhydroxy anthraquinone C-glycoside, carminic acid, which may be present at up to 20% of the dry weight of the mature insects. About 80,000 to 100,000 insects provide 1 kg of raw cochineal dye. Better quality and higher commercial value of cochineal correlate with a higher ratio of red to yellow pigments.

Cochineal pigments are extracted from dried bodies of female insects with water or with ethanol; the result is a red solution that is concentrated in order to obtain the 2-5% carminic acid required for commercial cochineal. For carmine lakes, the minimum content of carminic acid is 50%.

Cochineal extract or carminic acid is rarely used as a coloring material for food, but are usually offered in the form of their lakes. Aluminum (Al) complexes (lakes) can be prepared with various ratios of cochineal and Al varying from 8:1 to 2:1, having corresponding shades from pale yellow to violet. In solution at pH 4, carminic acid is yellow to orange, depending on concentration. At alkaline pH and in the presence of metals (mainly aluminium), it becomes bluish red. Carmines are defined as the lakes of carminic acid with various metals. The most used is carmine, the aluminium lake of carminic acid.

Cochineal carmine is insoluble in cold water, dilute acids, and alcohol, and slightly soluble in alkali giving a purplish-red solution. The shade becomes bluer at higher pH. Cochineal carmine is stable in light and heat, but the stability to SO_2 is poor.

In powdered form, carmine can be used for coloring various instant foodstuffs in alkaline solutions, and in ammonia for coloring different foods. Cochineal, carminic acid and carmines are approved as food colorants in the EU. The amount permitted in food ranges from 50 to 500 mg/kg. Carminic acid and carmine are considered very good food colorants due to their high stability and tinctorial properties. Solutions of carminic acid are yellow to orange, while carmines show various stable brilliant red hues. In the EU, the use of cochineal derivatives is authorized for coloring alcoholic and non-alcoholic drinks, candied fruits and vegetables, red fruit preserves, confectionery, ices, bakery products, cheeses, jam, jellies, marmalades, fruit-flavored cereals, and other products. It is currently the most robust red colorant among the authorized natural colorants, and its affinity for proteins means that it is an excellent colorant for dairy products. Cochineal added to processed pork meat provides a color similar to meat colored with erythrosine, but the color stability is higher. Cochineal can replace synthetic pigments (tartrazine, azorubine) for coloring jellies. In the US, cochineal derivatives are permitted in amounts ranging from 0.05 to 1.0%. The JECFA considered acceptable a total ADI of 0 to 5 mg/kg of body weight for carmines as ammonium, Ca, K, or Na salts.

Food Additive Regulations

The permission to use food colorants is bound to their safety and is strictly regulated by specific laws controlled at national and international levels. The natural colourants permitted by FSSAI for use in India and in US is listed in Tables respectively.

Table: Natural colourants permitted by FSSAI.

Color Type	Examples
Carotene and Carotenoids	β-carotene, β-apo-8'- carotenal, Methyl ester of β-apo-8'-carotenoic acid, Ethyl ester of β-apo-8'-carotenoic acid, Canthaxanthin.
Chlorophyll	
Riboflavin	
Caramel	
Annatto	Bixin, Norbixin
Saffron	
Curcumin or Turmeric	

Table: Application of natural colourants in food products as per US regulations.

Food Product	Colourant Used (Allowed)	Maximum Level (Ppm)
Malt bread	Caramel colours (E 150a-d)	NL
Breakfast cereals	Carminic acid Caramel color-III, Carotenes, Bixin, Norbixin, Capsanthin, Betanin, Anthocyanins	25-200
Chips (potato)	Curcumin, Turmeric	NL
Butter	Carrot extract	NL
Margarine, emulsions of fat in water	Curcumin, Turmeric	10 for E 160b

Cheese	Carotenes, Bixin, Norbixin	1.5-50 for E 160b
Vinegars and aromatic wines	Caramel colors (E 150a-d), Anthocyanins	NL
Bitter beverages	Curcumin, Turmeric, Riboflavin, Carminic acid	100 mg/lit.
Fruit and vegetable juices	Carotenes, Lycopene, β-apo-8-carotenal	NL
Beer and Cider	Caramel colors (E 150a-d)	NL
Vegetables conserved in vinegar, salt or oil	Riboflavin, Chlorophyll, Caramel colors (E 150a-d), Chlorophyllins, carotenes, Anthocyanins	NL
Jams and Marmalade	Curcumin, Turmeric, Chlorophyll, Chlorophyllins, Caramel colors (Ea-d), Capsanthin, Paprika, Lutein, Betanin, Anthocyanins, Carminic acid	100 only for synthetic colourants
Sausages, meat pastes, fish products	Curcumin, Turmeric, Riboflavin, Carminic acid, Caramel colors (Ea-d), Carotenes, Capsanthin, Paprika, Betanin	20-100
Hamburger	Carminic acid, Caramel colors (Ea–d)	100 only for E 120

NL – No limit specified.

Preparation of Colour Solutions

To prepare homogeneous solutions of water-soluble colorants, the powders or granules are dissolved in water or hydrophilic solvents (propylene glycol or glycerine), considering the solubility limits (1.5 - 4%) recommended by the manufacturer. To improve the stability of food colorants, specific permitted diluents and support matrices are used viz., lecithin, emulsifiers like polysorbates, monoglycerides and diglycerides, esters of organic acids with glycerin, esters of sucrose with fatty acids, sorbitan esters and viscosity regulating powders such as silicates.

Colorant Stability and Efficacy

The limiting factor for the efficacy of natural colourants is stability in the watery environments found in most food matrices. Generally the factors that deteriorate colorant stability and efficacy are physical (temperature, light), chemical (oxidizing or reducing agents, acids, alkalis), and biological (enzymes, microorganisms).

Application of Natural Colourants in Food Products

Natural Colorants in Beverages

Since inception, caramel colorants were used in Coca-Cola and Pepsi Cola. At present, natural colorants are becoming increasingly popular, especially in juice-containing beverages. The globalization of extraction and concentration techniques led to wider use of concentrated natural ingredients. They are easy to transport and to store and contribute to the popularity of naturally colored beverages in accordance with consumers' growing interest in quality and health protection.

Certain issues must be considered when using natural colourants in beverages. These include technical properties such as stability, sensitivity to light, pH, and temperature, and interactions with other ingredients and packaging. Some natural exempt colorants are used at concentrations of approximately 0.03 to 0.25% (annatto, cochineal, carmine, elderberry, grape skin extracts, caramel, β-carotene), others only as concentrates (turmeric and paprika oleoresins).

Lycopene can be used to color a variety of food products including beverages, dairy products, confectionery products, and baked goods. Tomat-O-Red (commercially produced lycopene extracted from non-GMO tomato) claims to provide health advantages (cancer preventing agent due to antioxidant potential) along with its function as a colorant.

When apples are pressed to yield juice, phloridzine, oxygen, and polyphenoloxidase enzyme combine to form phloridzine oxidation product (POP). This brilliant yellow natural colorant with nuances dependent on pH level can be incorporated easily into water-based foods such as beverages (juices, syrups) and confectionery creams, since it is stable during production processes.

Dairy Products Utilizing Natural Colorants

Dairy product companies are now producing beverages using flavonoids and anthocyanin-rich natural juices to color milk, whey-based and fermented products. Yogurt is generally colored with hydrophilic cochineal extracts, carmine, beet juice, or FD&C Red No. 3, all of which are resistant to microbial attack. Generally, the colorant is added to the fruit preparation, but must be stable at low pH levels. For example, beet juices are directly added to yogurt without the heat processing that may degrade the beet pigments. The most commonly used red colorants in yogurt are acid-proof cochineal extract and water-soluble carmine. Both are heat-stable at pH 2.8 to 7.0, and produce bright and appealing shades. Annatto emulsions showed less stability than annatto solutions and suspensions during continuous heating. GNT's Exberry product, is a new fruit and vegetable extract replacing annatto which is used as additive in Cheddar and Gouda cheese production.

Natural Colorants in Confections

Many sweets (confections) must be colored, a point in favour of their attractiveness for consumers. The commonly colored products are candies (starch jellies, candy cream centers, pan-coated candies, and hard candies), tablets, wafers, oil-based coatings, and chewing gums.

Natural colorants are rarely used for candy starch jellies due to their low stability to temperature and pH. Liquid colorants must be homogeneously incorporated into warm candy cream centers. Light discoloration is unacceptable and natural colorants are preferred. Red cochineal extracts (containing 3-4% carminic acid) are used at $pH > 4$ and turn orange at $pH < 4$, when added at the end of heat processing. Carmine lakes in alkali solutions are water soluble and may be incorporated in creams to yield bright magenta-red shades. The color can be changed by blending with yellow components (β-carotene or annatto). Beet juice may also provide red color but with lower heat stability and thus lower coloring power. Anthocyanins found in red cabbage, elderberry, black currant, grape juice, and grape seed concentrates are stable at $pH < 3.8$ and used for interior fillings but not for jellies due to their lower stability.

Orange shades are realized with lipophilic natural colorants like paprika oleoresin, β-carotene, and canthaxanthin after previous emulsification to yield water dispersible forms. Yellow shades can be achieved using turmeric as a water-soluble solution, but the solution is light sensitive. To maintain constant color, 3-6 ppm of β-carotene may be added. Stable brown coloration is obtained from caramel; concentrated syrup is easily incorporated, well flavored and stable in creams. Colorants must be introduced into the coating syrups during production of pancoated candies. Water-soluble colorants may be used but lake pigments as dispersions are preferred. Pan-coated candies require higher concentrations of colorants than jellies or creams; they require 30 to 60 coatings of colored syrup.

Opaque coatings are obtained by combining colorants and titanium oxide, and also using FD&C lake pigments. These dispersions may also contain stabilizers, preservatives, and viscosity regulators. The critical points are temperature (< 40 °C), dry matter (68 to 72%) and the ratio of tablet to coating syrup. Synthetic colorants are still preferred for coating because of better stability and lower cost, but the interest in natural colorants (turmeric, carmine, beet juice, β-carotene, red cabbage) in panned candies continues to increase rapidly.

Hard-boiled candies do not tolerate water after cooking, limiting the use of water-soluble colorants in vacuum cookers. The preferred method is to disperse these colorants into melted, hot candy slabs or disperse in glycerine or proplyene glycol. Fat candies mainly need oil-based dispersion coatings but may also accept water-soluble colorants. For gum products (sticks, balls), colored lakes blended with turmeric or carmine suspended in glycerine give bright colors.

Natural Colorants in Baked Goods

The use of natural versus synthetic colorants and water-soluble versus oil-soluble colorants depends on the cereal or flour matrix, baking temperature regime, pH, and ability of the colorant to dissolve in the appropriate solvent. Since baked foods contain fats, the colorants must be oil-dispersible, e.g., FD&C lakes or oil-soluble natural colorants. For a good dispersion in dough, water-soluble colorants are added at 1.5-3.0% concentration. Lakes are preferred for water-based coatings due to their better light stability and non-migration from the matrix. Liquid solvents such as propylene glycol or glycerine mixed with lecithin are recommended for cookie fillings because they incorporate well into vegetable oils.

The natural colorants used often for bakery products are annatto extract or annatto plus turmeric blends (0.02-0.06 %) to obtain yellow-orange shades. Crackers are colored with annatto extract, turmeric and paprika oleoresins, or caramel. Turmeric may be used also in combination with FD&C colorants.

Comparison of Natural with Synthetic Permitted Colourants for Food

Exempt colors need to be used at higher levels than their synthetic brethren. As a consequence, they may unexpectedly change the texture, odor, or flavor of the food. They are less stable and less consistent, causing the food at times to be unacceptable to the consumer. Exempt colors are often duller, more pastel, and more easily affected by the food matrix, pH, salts, vitamins, flavors, and other factors. They are also more likely to be contaminated with undesirable trace metals, insecticides, herbicides, and bacteria.

Epilogue

The present market for food colorants is estimated at 1 billion USD, while the natural food colorant market is only one-third of it. Synthetic colorants have achieved better results than natural or nature-identical colorants until now because of greater stability and higher ratios of coloring yield. In recent decades, the synthetic colorant market has declined, to the benefit of the natural-oriented market and consumers. In addition to the decreasing enthusiasm for chemicals in food, the high costs of toxicological studies also inhibit the development and approval of new synthetic colorants.

Synthetic Colours

Synthetic colourants or dyes are attractive to the food industry because they are superior to natural colourants in tinctorial power, consistence of strength, range and brilliance of shade, hue, stability, and ease of application. Synthetic colours provide a larger spectrum of colours. They also have lower prices and greater availability. A various food colorant with colour produced by them are given in Table.

Table: Synthetic Food Colourants.

Food Colourant	Colour
Allura Red AC	Yellowish red
Amaranth	Red
Azorubine (Carmoisine)	Red
Brilliant Blue FCF	Greenish blue
Brilliant Black BN	Black
Brown FK	Brown
Brown HT	Brown
Citrus Red No. 2	Red
Erythrosine	Bluish pink
Fast Green FCF	Bluish green
Fast Red E	Red
Green S	Green
Indigotine	Deep blue
Lithol Rubine BK	Deep red
Orange B	Orange
Patent Blue V	Blue
Ponceau 4R (Cochineal Red A)	Red
Red 2G	Red
Quinoline	Yellow Yellow
Sunset Yellow	Reddish yellow dye and lake
Tartrazine	Lemon yellow dye and lake

Previously, several different colourants were used in foods. However, there has been a decrease in the number of synthetic colourants permitted in many countries. Today, the types of colourants allowed vary greatly among countries, which reflect the different opinions about their toxicity.

However, there is a trend toward using fewer synthetic colourants. It is hoped that the trend to international standardization of food colourants will gain momentum. In order to prevent indiscriminate use many countries limit the types, uses, and amounts of colourants permitted in foods. Since different countries allow the use of specific food colourants, it is possible that foodstuffs may be imported into a country that forbids the colouring agent present in the product.

Therefore, methods capable of identifying and quantifying several colours simultaneously are desired in order to verify compliance to regulations. Information on the levels of these compounds in foods is also important to assess where the dietary intake stands compared to the ADI. According to the JECFA (Joint FAO/WHO Expert Committee on Food Additives), each country should verify periodically the dietary intake of colourants, and additives in general, to make sure that intake does not exceed the ADI.

Synthetic Food Colourants and their Uses as Food Additives

A various synthetic food colourants and their possible uses in food matrix as food additives are summarized below:

Colour

- Allura Red AC: ADI According to JECFA: 0 to 7 mg/kg bw. Utilization and Limits in Foods According to EU and US Regulation: Specific uses: bitter soda, bitter wine, other non-alcoholic flavored drinks alone or combined with other colourants (100 mg/l); luncheon meat (25 mg/kg), breakfast sausages with minimum cereal content of 6% (25 mg/kg); general uses: nonalcoholic flavored drinks (100 mg/l), candied fruits and vegetables (100 mg/l), red fruit preserves (200 mg/kg), confectionery (300 mg/kg), decorations and coatings (500 mg/kg), fine bakery wares (200 mg/kg), edible ices (150 mg/kg), flavored processed cheese (100 mg/kg), desserts including flavored milk products (150 mg/kg), sauces, seasonings, pickles, relishes, chutneys, and piccalillis (500 mg/kg), mustard (300 mg/kg), fish and crustacean pastes (100 mg/kg), precooked crustaceans (250 mg/kg), salmon substitutes (500 mg/kg), surimi (500 mg/kg), fish roe (300 mg/kg), smoked fish (100 mg/kg), extruded or expended snacks (200 mg/kg), other snacks (100 mg/kg), edible cheese rind (quantum satis), complete formula for weight control and nutritional supplements (50 mg/kg), liquid food supplement integrators (100 mg/kg), solid food supplement integrators (300 mg/kg), soups (50 mg/kg), meat and fish analogues based on vegetable proteins (100 mg/kg), other spirit beverages (200 mg/l), fruit wine, cider, perry, aromatized fruit wines (200 mg/l); FDA: can be safely used generally for colouring foods (including dietary supplements) in amounts consistent with good manufacturing practice; JECFA: 50 mg/kg limit in milk and 300 mg/kg in other foodstuffs.
- Amaranth: ADI According to JECFA: 0 to 0.5 mg/kg bw. Utilization and Limits in Foods According to EU and US Regulation: Aperitif wines and spirit drinks including products with less than 15% alcohol (30 mg/l); can be used in combination with other colourants, but not to exceed 100 mg/l; fish roe (30 mg/kg).
- Azorubine (Carmoisine): ADI According to JECFA: 0 to 4 mg/kg bw. Utilization and Limits in Foods According to EU and US Regulation: Specific uses: americano (50 mg/l), bitter and

wine, (50 mg/l); general uses: Non-alcoholic flavored drinks (50 mg/kg), candied fruits and vegetables, red fruit preserves, confectionery, decorations and coatings, fine bakery wares (50 mg/kg), edible ices (50 mg/kg), flavored processed cheese, desserts including flavored milk products (50 mg/kg), sauces, seasonings, pickles, relishes, chutneys, and piccalillis, mustard, fish and crustacean pastes, pre-cooked crustaceans, salmon substitutes, surimi, fish roe, smoked fish extruded or expended snacks, other snacks, edible cheese rind, complete formula for weight control and nutritional supplement, liquid food supplement integrators, solid food supplement integrators, soups, meat and fish analogues based on vegetable proteins, other spirit beverages, fruit wines, cider, perry, aromatized fruit wines; where not mentioned, max. level may not exceed amounts mentioned for Allura Red AC.

- Brilliant Blue FCF: ADI According to JECFA: 0 to 10 mg/kg bw. Utilization and Limits in Foods According to EU and US Regulation: Processed mushy and canned garden peas (20 mg/kg) and all foodstuffs and amounts mentioned for Allura Red general use; FDA: Can be safely used generally for colouring foods (including dietary supplements) in amounts consistent with GMP; JECFA: amount limited to 150 mg/kg in fermented milk and 100 mg/kg in baked goods.
- Brilliant Black BN: ADI According to JECFA: 0 to 1 mg/kg bw. Utilization and Limits in Foods According to EU and US Regulation: All foodstuffs and amounts mentioned for Allura Red general use.
- Brown FK: ADI According to JECFA: No ADI allocated. Utilization and Limits in Foods According to EU and US Regulation: Kippers (20 mg/kg).
- Brown HT: ADI According to JECFA: 0 to 1.5 mg/kg bw. Utilization and Limits in Foods According to EU and US Regulation: All foodstuffs and amounts mentioned for Azorubine general use.
- Citrus Red No. 2: ADI According to JECFA: Not to be used. Utilization and Limits in Foods According to EU and US Regulation: Permitted only for colouring skins of oranges, not intended for processing; max concentration is up to 2 ppm of whole fruit.
- Erythrosine: ADI According to JECFA: 0 to 0.1 mg/kg bw. Utilization and Limits in Foods According to EU and US Regulation: Cocktail and candied cherries (200 mg/kg), Bigarreaux cherries in syrup and in cocktails (150 mg/kg); FDA: Can be safely used generally for colouring foods (including dietary supplements) in amounts consistent with GMP; JECFA: can be used up to 300 mg/kg in various foods.
- Fast Green FCF: ADI According to JECFA: 0 to 25 mg/kg bw. Utilization and Limits in Foods According to EU and US Regulation: FDA can be safely used generally for colouring foods (including dietary supplements) in amounts consistent with GMP; JECFA: Can be used up to 100 mg/kg in various foods.
- Fast Red E No: ADI According to JECFA: No ADI allocated.
- Green S: ADI According to JECFA: No ADI allocated. Utilization and Limits in Foods According to EU and US Regulation: Specific uses: jam, jellies, marmalades, other similar fruit preparations including low-caloric products (100 mg/kg), processed mushy and

canned garden peas (10 mg/kg); can be used in all other foodstuffs in amounts mentioned for Allura Red general use.

- Indigotine: ADI According to JECFA: 0 to 5 mg/kg bw. Utilization and Limits in Foods According to EU and US Regulation: All foodstuffs and amounts mentioned for Allura Red general use. FDA: can be safely used generally for colouring foods (including dietary supplements) in amounts consistent with GMP; JECFA: Can be used up to 300 mg/kg in various foods.

- Lithol Rubine: ADI According to JECFA: No ADI allocated. Utilization and Limits in Foods According to EU and US Regulation: Edible cheese rind.

- Orange B: Utilization and Limits in Foods According to EU and US Regulation: Approved only in US; may be safely used only for colouring casings or surfaces of frankfurters and sausages, not more than 150 ppm by weight of finished food.

- Patent Blue: Utilization and Limits in Foods According to EU and US Regulation: All foodstuffs and amounts mentioned for Allura Red general use.

- Ponceau 4R (Cochineal Red A): ADI According to JECFA: 0 to 4 mg/kg bw. Utilization and Limits in Foods According to EU and US Regulation: Specific use americano (100 mg/l), bitter and wine (100 mg/l), jam, jellies, marmalades, similar fruit preparations including low-caloric products (100 mg/kg), chorizo sausage, salchichon (250 mg/kg), sobrasada (200 mg/kg); all foodstuffs and amounts mentioned for azorubine general use.

- Red 2G: ADI According to JECFA: 0 to 0.1 mg/kg bw. Utilization and Limits in Foods According to EU and US Regulation: Breakfast sausages with a minimum cereal content of 6% (20 mg/kg), burger meat.

- Quinoline Yellow: ADI According to JECFA: 0 to 10 mg/kg bw. Utilization and Limits in Foods According to EU and US Regulation: Specific use: Americano (100 mg/l), bitter soda and wine (100 mg/l), jams, jellies, marmalades, similar fruit preparations including low-caloric products (100 mg/kg); all foodstuffs and amounts mentioned for Allura Red general use.

- Sunset Yellow FCF: ADI According to JECFA: 0 to 2.5 mg/kg bw. Utilization and Limits in Foods According to EU and US Regulation: Specific uses: Bitter soda and wine (100 mg/l), jam, jellies, marmalades, similar fruit preparations including low-caloric products (100 mg/kg), sobrasada (135 mg/kg); all foodstuffs and amounts mentioned for Allura red general use; FDA: can be safely used generally for colouring foods (including dietary supplements) in amounts consistent with GMP; JECFA: can be used up to 300 mg/kg in various foods.

- Tartrazine: ADI According to JECFA: 0 to 7.5 mg/kg bw. Utilization and Limits in Foods According to EU and US Regulation: Specific use: americano (100 mg/l), bitter soda and wine (100 mg/l), processed mushy and canned garden peas (100 mg/kg); all foodstuffs and amounts mentioned for Allura Red general use; FDA: Can be safely used generally for colouring foods (including dietary supplements) in amounts consistent with GMP; JECFA: can be used up to 300 mg/kg in various foods.

ADI = acceptable daily intake, estimate of amount of a substance in food or drinking water, expressed as mg/kg body weight, that can be ingested daily over a lifetime without appreciable risk (weight of standard human = 60 kg); bw = body weight.

Synthetic Colour and Health

Food colours are being used in beverages, desserts, jams, jellies, sauces, pickles, cosmetics, toothpaste, etc. In addition, medicines, including tablets, capsules and syrups are dyed with food colours. Many of the food industries and other related companies as well as restaurants have relied on colourants in order to sell their products. However, there is a question on the safety and nature of colourants which have been used for these purposes.

Numerous studies have demonstrated the dangers of artificial colourants in food, which include the possibility of onset of attention deficit disorder (ADD), inhibition of the immune system, hyperactivity and allergic reactions. In addition, the use of non-permitted colours or overindulgence of permitted colours may also cause thyroid tumours, urticaria (hives) dermatitis, asthma, nasal congestion, abdominal pain, nausea, eczema, liver and kidney damage and cancer.

In view of the above, several food colourants have been banned in developed countries due to their toxicity observations on experimental animals. Table shows some of the widely used synthetic colourants and their toxic effects.

Table: Some of the routinely using food colourants and their effects.

Synthetic Food Colourant	Uses	Possible Effects
Quinoline yellow	Sweets, pickles	Asthma, hyperactivity, rashes
Ponceau 4R	Biscuits, drinks	Allergy, intolerance
Allura red	Soft drinks	Hypersensitivity
Azorubine	Sweets	Allergy, hyperactivity
Tartrazine	Sweets, biscuits	Asthma, hyperactivity, rashes
Sunset yellow	Ice creams, biscuits, Sweets	Gastric problem, allergy
Erythrosine	Toothpaste, Cough syrup	Hyperactivity, allergy

Food Flavoring

Flavoring or Flavouring is any of the liquid extracts, essences, and flavours that are added to foods to enhance their taste and aroma. Flavourings are prepared from essential oils, such as almond and lemon; from vanilla; from fresh fruits by expression; from ginger by extraction; from mixtures of essential oils and synthetic organic chemicals; or entirely from synthetic chemicals, with alcohol, glycerol, propylene glycol, alone or in combination, as solvents. Water is added and sometimes certified food colour as well.

Extracts, essences, and flavours employing only natural flavouring agents are called pure; those employing synthetics (in part or entirely) are called imitation, or artificial, flavourings. Essential oils, which are complex substances derived from plants, consist of a number of organic chemical components such as alcohols, aldehydes, ethers, esters, hydrocarbons (terpenes, sesquiterpenes, etc.), ketones, lactones, phenols, and phenol ethers. Nearly all of these organic chemicals have been synthesized, and it is these synthetics that are used in the manufacture of imitation flavourings.

Essential-Oil Extracts

These extracts are prepared by dissolving an essential oil in alcohol of the proper strength, adding water and, where desirable and permitted by law, a small amount of certified food colour. They include almond, anise, celery, cassia or cinnamon, clove, lemon, nutmeg, orange, rosemary, savory, basil, sweet marjoram, thyme, and wintergreen.

Non-Alcoholic Flavouring Extracts

Non-alcoholic flavouring extracts are prepared by employing glycerol or propylene glycol or a combination of these to bring the essential oil into solution, with added water, and sometimes food colour.

True Fruit Flavours

True fruit flavours are obtained by expression, concentration, or distillation of fresh fruits. The alcoholic content of the finished product is usually between 18 and 22 percent to prevent fermentation.

Imitation, Artificial Extracts, Essences and Flavours

Imitation, artificial extracts, essences, and flavours are prepared by bringing into solution with alcohol, glycerol, or propylene glycol various synthetic flavouring agents to formulate an extract, essence, or flavour with the likeness of the flavour of the fruit, spirit, or liqueur for which it is named. These preparations cover a wide range of flavourings including vanilla, lemon, lime, banana, cherry, butterscotch, brandy, and rum. Some contain few ingredients, others many. Imitation vanilla extract may be prepared with synthetic vanillin, coumarin (now prohibited in the United States and Canada), heliotropin, glycerol, water, and caramel colouring.

Imitation strawberry flavour may contain 10, 12, or more synthetic organic chemicals including aldehyde C16, aldehyde C14, benzyl acetate, methyl anthranilate, amyl acetate, amyl butyrate, and ethyl propionate, with solvents and added colour, according to the particular manufacturer's art, skill, and experience in bringing out the fullest and most realistic flavour.

Imitation brandy flavour may contain, with other organic chemicals, ethyl oenanthate, methyl succinate, ethyl acetate, and ethyl propionate. Certain synthetic organic chemicals lend a distinctive note to imitation flavourings for example, allyl caproate and ethyl butyrate to pineapple, benzaldehyde to almond, benzyl butyrate to raspberry, and citral to lemon.

Functions of Flavours

The role of a flavour is simply to impart sensory pleasure to a good, beverage, tobacco, or pharmaceutical, and in doing so it serves a diverse function. It is important to understand the function expected of the flavour so that the type best suited for this function can be delivered. As differentiated in Table flavour has three major categories of function: Economic, Physiological, and Psychological. Many flavour applications perform functions in all three categories, although each category will have a different emphasis. For example, some applications are mainly for pleasure (psychological functions) as exemplified by a hard-boiled candy as opposed to a flavoured yogurt (for breakfast) where the three functions economic, physiological, and psychological are applied in differing degrees.

Table: Functional uses of flavours.

Economic	Physiological	Psychological
Simulate	Metabolic response	Nostalgia
Extend	Intestinal absorption	Association
Flavour the unflavoured	Appetite and consumption	Intellect/belief, cognitive factors
Modify (cover taste)		Trend
Compensate for flavour losses		Flavour the flavoured
Improve shelf-life		

- Economic Functions: Some healthful, nutritional food preparations might have undesirable tastes (e.g., soya and vitamins). Flavours that can modify their taste and make these nutritional, economical food preparations more palatable would be desirable. Another example of this application is the use of flavour to compensate for flavour losses in food processing or to extend freshness during storage. One important function conceptualized by Givaudan Roure is to “flavour the unflavoured” which is used to bring variety and palatability to a bland food such as rice or cereal.

- Physiological Function: Various studies on taste and fat digestion provide some indication that taste can alter the metabolic response to a fatty meal. There is also research on taste and intestinal absorption of glucose that proposes oral stimulation affecting intestinal absorption. However research in this regard is still in infancy stage.

- Psychological Functions: Although the main role of a flavour is to provide sensory pleasure, psychological analysis of this pleasure can greatly assist in proper flavour selection. Examples of today‘s trendy flavours are the fantasy types that do not necessarily represent a commonly known food, but rather a novel combination, thus creating a new identity. These are exemplified by the various fruit punches, candies, toffees, etc. The concept of “flavour the flavoured” is also useful to add variety and increase sensory pleasure to foods that normally need no further flavouring. The addition of flavours to tea and coffee is an example of this concept.

Classification of Food Flavours and Uses

Food Flavours are classified into three major categories:

Natural Flavours

- Herbs- Basil, mint.

- Spices- Cardamon, clove, turmeric.
- Aromatic Seeds-Anissed, Cumin.
- Fruits- Orange, Lemon.
- Vegetables- Pees, Onions, Garlic.

Processed Flavour

- Caramelized
- Roasted
- Fermented
- Toasted
- Baked

Added Flavour

- Natural Extracted Flavour:
    - Essential Oil
    - Essence
    - Extracts
- Synthetic Flavour:
    - Fruit Flavour
    - Savoury Flavour

Natural Flavours

- They are usually extremely complex mixtures of many different substances. Sometimes the flavour of natural flavouring agent may depend upon a single substance. Eg:- Clove oil flavour is because of chemical eugenol. It contributes 85% of clove oil.
- Or it may be present in extremely small amounts such as citral in oil of lemon which constitutes 5% of the oil.
- Natural flavouring agents are compound of mainly aromatic organic compounds present as volatile essential oils or as non-volatile constituents. Such as resins and oleoresins. They are formed in the plant during normal plant metabolism and remains as such the plant is harvested.
- The aroma of onion, garlic, cabbage is mainly due to Sulphur-containing compounds.

- The flavour and aroma of fruits and vegetables are because of the presence of volatile organic chemicals such as esters, aldehydes, acids, alcohols, ketones and ether present as essential oils in natural foods.

Processed Flavour

- The flavour that develops during processing by decompositions, the combination with other compounds, or formation of a new compound.
- The following flavours form during processing of various foods:
    - Flavour due to enzyme action.
        - Vegetable cuts (onions, garlic). Odour is because of formation of diallyl disulfide.
    - Flavour produced by microbiological action during fermentation of sugar.

Fermentation

Eg- sugar —–-yeast—–>>alcohol

Milk——bacteria——>>curd

Milk——bacteria—–>>cheese

- Flavour formed during cooking and other heat processing. Eg: cooked meat flavour, aroma of freshly baked bread or roasted Coffee beans.
- Undesirable flavours caused oxidation eg: rancidity of oil.

Added Flavour

- Natural flavour or synthetic flavours are often added to food to increase its acceptability.
- Role of added flavours has gained importance with advancement in technology being used in food industry and development of new products such as bakery and confectionary items, ready to eat foods, beverage ad fast food items.
- The flavour which is lost during cooking and processing can be replaced by either of the following methods:
    - Adding natural flavourings and extracts:– Eg:- Natural essence from fruits, essential oils extracted from spices, Vanilla essence is prepared by extracting the essential oil from vanilla pod with the channel.
    - Adding synthetic flavours that consist of a blend of chemical which smells like the original substances. Eg:- Vanillin is the synthetic flavour instead of vanilla.
- Synthetic chemicals are blended to match a natural flavour.
- Other substances which contribute towards the flavour of food are sweeteners and flavour enhances.

Sweeteners

Sweeteners used can be:

- Natural Sweeteners:
  - Carbohydrate sweeteners,
  - Nutritive Sweeteners.
- Artificial Sweeteners:
  - Synthetic Sweeteners,
  - Non-Nutritive,
  - They do not provide calories.

Sugar is added to savoury dishes in minute amounts to improve the flavour of the dish.

Flavour Enhancer

- They are chemicals which by themselves have little or no odour or taste.
- When they are added to food, even in small quantities. They bring out flavour of the food.
- That us they are capable of enhancing, modifying, or intensifying the original flavour.

The following chemicals are used as flavour enhancers:

Monosodium Glutamate(MSG)

- Also called as aji-no-moto or Chinese salt.
- It is sodium salt of glutamic acid.
- Glutamic acid is present in many foods, its content decreases during harvesting to processing to processing causing a natural loss of flavour.
- This can be partly restored by adding MSG to food.
- It draws out the hidden flavour of food and reduces the less desirable flavours.
- MSG increases salivation and its excessive consumption leads to a condition known as "Chinese restaurant Syndrome" in which people suffer from various symptoms such as a burning sensation, migraine like symptoms and chest pain.
- It is used a flavour intensifier in soups, sauces, gravies, taste maker and flavourings.

Nucleotides

- They are used as flavour enhancers.

- They are 500 to 100 times stronger than MSG.
- MSG & nucleotides are ideally used together in the ratio of 1:50.
- 1g of nucleotides is used with 50g of MSG.
- It is used in processed foods such as potato, chips, peanuts, dry and canned soups, sauces, ketchups etc.

Maltol

- Used as a flavour enhancer for sweet flavours.
- It is found in several plants and is formed when cocoa, coffee, and malt are roasted and in bread when it is baked.
- It is synthesized from soybean proteins fermentation and used as a fragrant, caramel flavour addition of fruit based products.
- Also used in cookies, beverages, in instant pudding mix.

Salt

- It is used on food for its flavour, as a preservative and as a dietary constituent.
- Main role is for salty taste, flavour intensification and as a digestive stimulant.

Sodium Restricted Flavouring

- Herbs, spices like pepper, dry mustard, paprika, used when salt is restricted in diet.
- Salt substances are salt which does not contain sodium but contain potassium or ammonium.

Spices and herbs contribute natural flavour to foods in which they are used or added to. They are used to enhance the flavor of the food. The flavour is due to essential oils and other flavour components present in the plants. Herbs are the leaves of fresh or dried plants, while spices may be the aromatic part of plants, usually dried buds, fruits roots, bark. They both have nutritive and medicinal properties and are used in cuisines all over the world. Common herbs- basil, bay leaf, curry leaves, mint, oregano, parsley, thyme. Common spices- ajwain, asafoetida, fenugreek seeds, mango powder, turmeric.

Use of Flavour in Food Preparations

- Retaining the natural flavour while processing is an art. An understanding of volatile and non-volatile flavours is necessary for a chef.
- Flavours are often lost when food becomes stale, and off flavour develop.
- Processing of food results in both developments of flavours and loss of flavour.

- Different processes and ingredients used give different flavours to food.

Flavourings agents are widely used in the food industry to:

- Enhance a good flavour eg: addition of chocolate essences to chocolate cake.
- Replace flavour lost while processing food.
- Give particular flavour to food.
- Mask some undesirable flavour in order to increase acceptability of food.

Eg: Adding essence to egg not to mask the egg flavour.

Food stuff	Flavouring agent
Mint	Menthol
Thyme	Thymol
Cloves	Eugenol
Pepper	Piperidine
Butter	Diacetal
Orange	Limonene
Lemon	Citral
Garlic	Diallyl disulfide
Turmeric	Curcumin

Flavor in Different Foods

- Tannin - A mixture of strong astringent acids found in plants, particularly tea leaves, red Grape skins and the bark of trees. It has the ability to coagulate proteins and is Responsible for the keeping quality of fine red wines.
- Tannic acid - One of the acids in tannin used for flavoring and as a clarifying agent in beer, Wine, cider, and other brewed drinks.
- Caffeine - Coffee owes its characteristic flavor to caffeine although by itself caffeine without Its aroma has a faint bitter taste. Coffee also contains alkaloids, volatile aromatic Products and substances belonging to the phenolic series. It stimulates the central Nervous system. The composition of coffee is – 15.30%-nitrogenous substances, 11.40%-fatty matter, 70-2%-caffeine.
- Capsaicin - Contained in chilies and is responsible for their fiery flavour. It is an oily Substance insoluble in water. The heat of the chilly is measured in Sackville units.
- Sin grin – Is present in raw cabbage, which in the presence of heat is converted to Isothiocyanate and ultimately to hydrogen supplied- a strong unpleasant smelling Compound. Therefore cabbage should be cooked for a short time, covered with a Lid.
- Alcohol - Wine has alcohol which refers to a potable liquid containing ethyl Alcohol

(C_6H_5OH) it denotes a class of organic compounds distinguished by the presence of hydroxyl group. Alcohol is produced as a result of fermentation.

Condiments

Condiments are aromatic substances added to food to improve its flavor. The term "seasoning" is applied to substances which are added at the time of cooking while "condiments" apply to those added at the table to food already prepared. Condiments are classified according to their dominant flavors:

- Acid - Vinegar, lemon juice.
- Bitter (aromatic) - Pepper, paprika.
- Bitter - Garlic, shallot, wheels onion, spring onion, leeks.
- Fat - Oil, butter, fats.
- Readymade condiments – English sauces (Worcestershire, Harvey) ketchups, curry Powders, prepared mustards, soy sauce etc. 6. Salt condiments-sodium chloride, sea salt.
- Sweet condiments - Sugar, honey.

Other Flavouring Categories

- Flavouring Preparations: Flavouring preparations are not chemically defined substances but complex mixtures whose composition is defined by natural raw materials. Well known examples are vegetable and fruit extracts, spice and herb extracts and yeast extract. As is the case with natural flavouring substances, flavouring preparations are obtained from plant, animal or microbiological source materials by means of physical or biotechnological production processes. Flavouring preparations are often the main component of citrus, spice and mint flavourings, reflecting their characteristic smell and taste. This category also includes essential oils, such as clove and eucalyptus oil.
- Thermal Process Flavourings: Thermal process flavourings are industrially produced by the controlled heating of several components (e.g. by a Maillard reaction). During the process, intense flavours develop, as is the case when roasting meat or baking bread. The flavouring itself only develops as a result of the thermal process and the essential basic materials for their development include amino acids and also reducing sugars, such as dextrose (glucose).
- Smoke Flavourings: Smoking is one of the most ancient procedures to season and preserve foods. Smoke flavourings are not only used to preserve but to confer a special smoke flavour in foods such as fish and meat products, steak, sauces and chips.

  Freshly produced smoke is obtained by means of controlled burning of hardwood at temperatures of up to 600 °C. This happens mostly in the absence of air, and in many cases charcoal develops as a by-product. The smoke generated is condensed and for practical reasons, such smoke primary products can be mixed with solvents such as cooking oil or carriers such as table salt.

Table: Source of raw materials used in flavouring compounds.

Synthetic	Natural	
	Botanical	Animal
Benzaldehyde	Fruit and vegetable juice, extract and distillate.	Plasma drippings
Cinnamic Alcohol	Herbs	Seafood by-products, en-zyme- modified cheese, meat extract.
	Spices	
	Nuts	

Table: Classification of Nature's Flavours.

Source	Primary Origin	Secondary Origin	
		Biological	Thermal
Botanical	Fruits, vegetables, spices, flowers, nuts	Wine, beer, bread	Coffee, cocoa, caramel
Animal	Fish, beef, chicken, milk	Fermented sausage, cheese	Roast beef, boiled chicken, grilled cheese

Table: Organic Synthetic Chemicals Used in Flavours.

Group	Subgroup	Example of chemicals	FEMA No.	Natural Occurance	Organoleptic Characteristics
Aromatic-Benzenoid	Phenols	Eugenol	2467	Clove oil, banana, cinnamon, cocoa, coffee	Clove-like, spicy
	Ethers	Anethole	2086	Anise, fennel, basil, mint, cheese, tea	Anise odour, sweet herbaceous
	Acetals	Acetal	2002	Apple, grape, bread, whisky, rum	Fruity, green
	Carbonyls	Acetanisole	2005	Anise seed, tomato, tea	Floral, bitter
	Carboxylic acids	-	-	-	-
	Esters	Methyl	2745	Apple, cherry, tomato,	Characteristic,
		Salicylate		wine, winter green oil	wintergreen
	Lactones	Dihydrocoumain	2381	Sweet clover	Spicy, vanilla
	Sulfur com-pounds	5-methyl-2- thio-phene-carboxaldehyde	3209	Roasted peanuts	Strong, nutty, meaty

Aromatic-Heterocyclic ring	Furans	Furfural Mercaptan	2493	Coffee, beef	Strong, unpleasant
	Pyrans	Maltol	2656	Roasted malt, chicory, strawberry, bread	Coffee-like, sweet, fruity, jam-like
	Thiophenes	-	-	-	-
	Pyrazines	2,3,5-Trimethyl pyrazine	3244	Baked goods, coffee, cocoa, peanuts	Sweet, roasted peanut
	Imidazoles	-	-	-	-
	Pyridines	Pyridine	2966	Wood oil, coffee, tobacco	Penetrating, fishy odour, burnt
	Pyrroles	2-acetyl pyrrol	3202	Bread, cheese, tobacco, tea	Strong, roasted
	Oxazoles	-	-	-	-
	Thiazoles	2,3,5-Trimethyl thiazole	3325	Potatoes, beef, coffee	Chocolate, nutty, coffee
Aliphatic-Cyclic	Lactones	γ-Decalactone	2360	Peach, apricot, butter, cheese, meat	Pleasant, fruity, peach-like, creamy
Aliphatic - Acyclic	Hydrocarbons	d-limonene	2633	Lemon, orange, mandarin, peppermint	Weak orange or lemon
	Alcohols	Cis-3-hexenol	2563	Apple, orange, raspberry, grapefruit	Intense, green odour, leafy
	Carbonyls a. ketones b. aldehydes	a. Octanal	2797	Orange, mandarin, grapefruit, rose, beef	Fatty, orange
		b. Acetaldehyde	2003	Fruits, tobacco, orange, nuts	Pungent and penetrating
	Carboxylic acids	Butyric acid	2321	Dairy products, bread, strawberry, beef	Rancid, sour milk
	Esters	Ethyl butyrate	2427	Strawberry, olive oil, apple, cheese	Fruity, powerful
	Isoprenoids	Citronellal	2307	Citronella, lemon, mandarin, grape	Floral, citronella, rose-like
	Sulfur compounds	Methyl sulphide	2746	Dairy products, meat peppermint	Unpleasant, cabbage-like
		Allyl disulphide	2028	Garlic, meat, onion	Characteristic garlic, pungent
		Methyl mercaptan	2716	Meat, cheese, bread	Objectionable, rotting cabbage
	Nitrogen compounds	Piperidine	2908	Black, pepper, tobacco, bread, meat, fish	Heavy, sweet, animal like

Process of Flavour Creation

The oldest and simplest method for flavour creation is the artistic approach. The second approach is to combine art with scientific know-how. The third approach is to follow nature's footsteps and develop flavours bio-synthetically.

The Artistic Approach

In this approach, the artistic powers of the flavour chemist are at work. These several components comprise. The sensory characteristics of flavour components like synthetic organic chemicals, natural essential oils, botanical extracts, juices, spices, nuts, herbs, and concentrates of animal origin, leave various impressions. Each flavour chemist will have the characteristics of these components and their corresponding food-nuance associations stored in memory, dependent on his or her personal ability and training.

This approach can be called an iterative method or a trial by error approach. The flavourist produces a trial formulation, evaluates the trial by taste, and then makes revisions based upon their evaluation. This is the traditional type of methodology and has been used by flavourists for many years.

The Scientific Approach

The artistic approach is used along with the scientific knowledge. The utilization of chromatography, along with various other analytical techniques, has the flavour chemist gain a better understanding of nature's process in producing flavours. This knowledge is used in combination with the artistic skills of the flavour chemist to create unique flavours.

The Biosynthetic Approach

In this approach, the biochemist attempts to duplicate nature's biogenetic pathways. Although science has not yet unlocked many of nature's secrets in developing its flavours, some of the known enzymatic and fermentation reactions are crudely exploited to produce building blocks. Enzymatically modified cheeses, fermented fruits and wine, and cooked and roasted foods are just a few examples. Due to the increased demand for natural flavours, research in this area is dramatically increasing. The primary areas of biotechnology that are being utilized and explored for use in the flavour business are:

- The product of natural flavour ingredients.
- The production of part or complete flavours.
- Increased yields of essential oils, oleoresin, and flavour components by the use of enzymes in the processing of natural materials.
- Plant cloning for maximum yields of secondary metabolites.
- Specific chemical steps to produce economies in the production of expensive flavour chemicals or to support natural claims (e.g., production of natural benzaldehyde from other natural chemicals).

- Modified traditional fermentation systems to produce flavour enhanced bases for direct use or further concentration.
- Secondary metabolites from in vitro tissue culture of vegetable cells.

Table: Examples of Biosynthetic Production of Chemicals.

Sr. No.	Major product(s) pro duced in nature	Examples of occurrence	Microorganism applied*	Major portion of substrate
1	Methyl ketones	Cheese	Penicillium	Fatty acids
			Roqueforti	
2	Lactones	Peaches, coconut	Pityrosporum species	Lipids
3	Butyric acid	Butter	Clostridium Butyricum	Dextrose
4	Carveol, carvone, di-hydrocarvone, perillyl alcohol	Spearmint and other essen-tial oils	Pseudomonas species	Limonene
5	d-Verbenone, d-cis-Verbenol	-	Aspergillus Niger	α-Pinene
6	Cheese like flavour	Cheese	Streptococcus species Lactobacillus species	Reconstituted milk
7	Bread like flavour	Bread	Saccharomyces Cerevisiae	Sugar and milk

* Exclusively food microorganisms are applied.

Thermally Produced Flavours

In this process, the chemist follows the natural route of forming the "secondary origin" flavours mentioned in Table. Ingredients used in their preparation consist of one or more of the following:

- A protein nitrogen source, for example, meat, poultry, dairy products, seafood, and their hydrolysis products.
- A carbohydrate source, for example, vegetable, fruit, sugars, and their hydrolysis products.
- A fat or fatty source.
- Herbs, spices, water, selected vitamins, acid, emulsifiers, and nucleotides.
- Flavouring preparations and flavour adjuncts to be added after processing.

The prepared mixture is then processed under controlled conditions. The temperature should not exceed 180 °C for the specified time. The general concept of producing flavours thermally is to utilize foodstuffs or constituents of foodstuffs and apply processes comparable to traditional kitchen treatments.

Recent Flavour Creation Technologies

Flavour companies are using computer technology to assist the flavourist in their creative task. Many companies have introduced systems where a flavourist has access to a database of thousands of different flavour preparations. By doing selective searches through this database, the flavourist can retrieve a starting formulation which meets specific parameters (e.g., liquid, natural, kosher, and heat stable). From there, the flavourist can tailor a new formulation that meets a specific customer's needs. By the use of this type of technology, the time to develop a new flavour has been reduced significantly. There are also equipments developed, which allows flavourist to evaluate almost instantaneously the aroma of blends they make using computer technology even to the extent of including any aroma of a food product—for example, yogurt. Other equipment is used to accurately measure threshold levels either of individual components or of flavours in finished food products. By using this type of equipment, the flavourist, supported by sensory specialists, can evaluate the different release rates of a flavour in different bases or products and then can adjust formulations so that the same flavour perception can be achieved in different applications.

General Rules for Flavour Creation

In spite of the increasing number of technological processes and tools, artistry is still the most important factor in the process of flavour creation. Modern flavour chemists utilize all the scientific resources available to them to create the proper balance between art and science to achieve excellence. Attempts to use science only, such as using analytical tools to recreate flavours from data produced, have proven to be fruitless. Reasons for failure of such an approach can be summarized as follows:

- Only a few of the hundreds of chemicals identified in foods have sensory significance.
- Availability and cost of the total chemicals identified will prove prohibitively costly.
- The methodology to obtain qualitative and quantitative identification of trace chemicals is not very accurate.
- Synthetic chemicals, utilized to duplicate nature, contain trace impurities that affect the final taste of the compound.

Therefore, technical resources are tools that should be employed to service the flavour chemist.

Important factors influencing flavour creation are:

- Flavour Profile of the Target: The flavour target could be any flavour, aroma fruit such as a mango to roast beef. Drawing the preliminary flavour profile of this target is an important step in the process of the flavour development and constitutes the foundation upon which the flavour will be created.
- Flavour Descriptive Language: To reduce communication gap between industry, customers and also amongst flavour chemists for description of a particular flavour, it is imperative to use a standard descriptive language among flavour chemists and food scientists.
- Sensory Evaluation: Application of proper sensory techniques and large panels will make

the difference in the fine tuning process. Reviewing the flavour profile by utilizing focus groups will bring the right balance of art and science. Consumers should be the ultimate judges of the quality of a flavour.

- Reproducibility: Since flavours are complex mixtures of various components, the flavour chemists' task is to ensure the compatibility of such mixtures and also it is important that any laboratory process should be reproducible in production.

Analytical Techniques for Determining Flavour Components

The process of determining the flavour components in a food can be divided into three general phases.

1. Isolation of the volatile flavour components from the bulk of the non-volatile matrix: Once the sample has been reduced to the proper physical form, the volatile flavour components are usually isolated by either distillation or extraction, or a combination of the two. Since most liquid samples are in an aqueous medium, simple extractions with an organic solvent (either polar solvents such as methylene chloride or nonpolar solvents such as ether or pentane) in a separatory funnel are used. There are also solvents that can be used in special cases, such as isopentane, which can extract the volatile components from a matrix high in ethanol or propylene glycol. soxhlet extraction apparatus is used for the extraction of solid samples, such as ground nuts or seeds. Since this method collects the extracted volatiles in boiling solvent, heat-sensitive compounds may be lost or produce artifacts. Where significant heat-sensitive materials are present, an extraction in which solvent is continuously recirculated through the sample by means of a pump at ambient temperature is more suitable.

2. Separation of the isolated components into individual or small groups of components: Column and thin-layer chromatography are generally used for rough separations into groups of components, while gas chromatography (GC) and high-performance liquid chromatography (HPLC) are of greater value in determining individual components.

3. Identification of the flavour components: There are a wide variety of tools to identify unknown components. If the general class of the flavour components is known and compared to an accumulated library of standards, retention time data can be used for identification. Tables of retention indices such as the Kovats index have been published. Kovats indices provide the retention times of individual components compared with those of a homologous series of alkanes or esters. SNIFF analysis may also be used. In this method, an unknown sample is injected into the gas chromatograph and is separated into the various components. As the components are eluted, a portion of each is sent to the detector and a separate portion is also sent to a sniffing port where the flavourist evaluates the odor of the component as it is released from the gas chromatograph and identifies the component. Mass spectrometry (MS) is also a widely used tool to identify unknown components. The mass spectrometer is attached to a gas chromatograph which separates the unknown sample into its components. With the progress in computers GC-MS systems are nowadays available at reasonable costs.

Applications of Flavours

Flavours are applied to various products in the food, beverage, tobacco, pharmaceutical, and oral hygiene areas. Lately, some applications have extended to other segments such as the toy industry. Flavoured products fall into two categories:

- Flavour-dependent: These are foods and beverages that cannot exist without the application of flavours. Examples are hard-boiled candy, chewing gum, carbonated and non-juice drinks, gelatin desserts, and powdered artificial beverages.
- Flavour-independent: These are products that can be marketed without flavours (e.g. crackers, cereals, and nuts) or for which flavours are legally prohibited (e.g. milk, orange juice, and butter wherein flavour reinforcement is not permitted, unless a new identity is given to the food).

Direct and Indirect Flavouring

It is interesting to note that almost every food can be flavoured provided that consumers are not misled or deceived. The food industry achieves this, as in the following examples:

- New identity: A food such as milk, butter, or juice can be flavoured provided that it is name to convey to consumers its new identity and avoid consumer deception. Examples are blends of butter and margarine.
- Indirect flavouring: Indirect flavourings are preparations that might be added by consumers to their home-cooked meals e.g. Salad dressings.

Flavour Selection

It is important for flavour users to understand the complexity of flavour selection. It will be helpful to begin by separating flavours from all other additives and placing them in a class by themselves. The main reason for the said purpose is Specificity. Flavours are very specific to certain applications. For example, the same orange flavour (same identification number and supplier) might taste excellent in some applications but be much less acceptable in others. Flavours in many instances are tailored to suit a complex food system as well as the processing conditions applied.

Flavour Forms

Flavour strength and potency vary considerably. Some flavours are diluted with solvents and carriers, whereas others are compounds of aromatic chemicals without a solvent. Various flavour forms are delineated in table.

Table: Forms of Flavours available in Market.

Type	Forms	Solvents and carriers
Liquid	Water, alcohol or oil-soluble	Alcohol, propylene glycol, triacetin, benzyl alcohol, glycerin, syrup, water, vegetable oil
Powder	Spray-dried, absorbates, or powder mixes	Gum acacia, starch hydrolysates, selective hydrocolloids, simple carbohydrates
Pastes and emulsions	Emulsion of the oil-in-water type	Same ingredients as for powder and liquid

Flavour Regulations

The purpose of food laws is to assure safety for the consumer and prevent deception. Labeling regulations require that information be provided on the kind of food, its processing, and the additives contained in it. These principles also apply to flavours. The vast number of flavouring raw materials makes this a difficult task for legislators, resulting in various countries taking different approaches. The purpose of flavouring a food is to add sensory value (pleasure) to that food without altering its identity. Most food regulations are formulated to put a stop to practices where flavour might be used to give false values. The addition of a characterizing flavour is usually not permitted to compensate for the lack of natural flavour due to poor or improper processing. For example, butter or cheese flavours are not permitted to flavour natural butter or cheese, respectively.

Flavour Safety

Flavourings are a part of food, and they must be safe for human consumption. Only regulated ingredients that are recognized to be safe are permitted. Types of systems regulating addition of flavour components in foods are:

- Positive List System: A positive list is composed of flavouring raw materials that are believed to be innocuous and safe for use in food. Positive lists are used mostly to regulate the use of food additives such as preservatives, colorants, and antioxidants.
- Negative List System: A negative list records materials that shall not be used in flavourings because they are known to be harmful and also lists substances for which an upper consumption limit has been fixed.
- Mixed System: The mixed system was introduced in some European countries to combine the advantages of the positive and negative list systems. The advantage of the mixed system is the clear listing of forbidden substances, providing adequate consumer protection; it can be enforced in practice without undue cost.

The Flavour and Extract Manufacturers Association (FEMA) of the United States was founded in 1909 by several flavour firms in response to the passage of the Pure Food and Drug Act of 1906, through its "generally recognized as safe" (GRAS) assessment of flavouring substances. FEMA is also a founding member of the International Organization of the Flavour Industry (IOFI), which is an association of regional and national associations of the flavour industry, headquartered in Switzerland.

Labeling

Flavours are declared in finished food products according to national food labeling regulations. The labeling of foods in relation to added flavours is regulated in many countries. The purpose is to label the flavours in conformity with their true nature so that consumers will be properly advised as to the type of flavours used. Although regulations differ greatly from one country to another, the chief objective is to differentiate among the main types of flavours:

- Natural flavours

- Nature-identical flavours (where applicable)
- Artificial flavours
- Mixtures of natural flavours and nature-identical or artificial flavours

This is a simplification of terminology. What is considered natural or artificial in one country is not necessarily perceived as such in another.

Natural Flavors Obtained by Microbiological Pathway

Generally, the first major source of flavor is the extraction from plant biomass due to consumer preference for "clean" and "organically" produced aromas and fragrances. Taken into account the significant differences between the price of synthetic and non-synthetic manufacturing, the microbial flavour production is considered.

Lactic acid bacteria (LABs) are an important class of microorganism for flavour manufacturer. LABs are very important for the dairy industry, being extensively used in fermented food production. During the fermentative processes, LABs influences the sensory properties of the final products, also including the flavour development. The flavour production is very much substrate and strain dependent, and the presence of the flavour precursors and regulatory responses may influence the balance of the flavour biosynthesis from a secondary metabolite product to the main compound. Strains like Lactococcus lactis subsp. lactis and Lactococcus lactis subsp. lactis var. diacetylactis are industrially used for flavour biosynthesis as a sole microorganism or in coculture with Streptococcus thermophilus and Lactobacillus bulgaricus.

At the same time, the flavour biosynthesis and the changes in metabolic pathway are linked to environmental conditions. From a technological point of view, the metabolic changes are very important for the volatile compound biosynthesis, as well as for the microorganism, in order to obtain energy and to maintain the $NAD^+/NADH^+H^+$ balance. It is obvious that there are major differences in flavour profiles between utilised complex and standard media. S. thermophilus LMG18311 biosynthesize 2,3-pentanedione and acetic acid only in standard media, but Bacillus subtilis CICC 10025 biosynthesize higher amounts of acetoin in media consisting of acidified molasses and soybean hydrolysate, because soybean hydrolysate is a more optimal nitrogen source for acetoin production for this strain. Diacetyl is almost exclusively synthesised by LAB and is the key flavour compound naturally produced by the Leuconostoc sp.

The Influence of the Culture Medium Composition on Flavour Biosynthesis

The Influence of Nitrogen

The LAB strains are able to survive starvation due to their capacity to utilise another energy source rather than carbon. The starvation conditions decrease the organism ability to synthesise ATP with generation of proton motive force (PMF) and also slow down the accumulation of necessary nutrients to maintain viability over time. As an additionally carbon source, the LABs are capable to catabolise amino acids which provide building blocks, cofactor recycling and limited energy source. The LAB inability to synthesise many of the amino acids required for protein synthesis needs the supplementation of the culture media with high amount of essential amino acids, since the amino acid catabolism is a major process for flavour formation. Proteolytic enzymes from LAB play an important role in degradation of proteins by producing free amino acids. These amino acids contribute directly to flavour formation being precursors for catabolic reactions. The conversion of amino acids to aroma compounds by LAB is essentially initiated by a transamination reaction, which requires α-ketoacid as the amino group acceptor, pathway demonstrated for lactococci, mesophilic lactobacilli and thermophilic LAB.

By amino acid catabolism, LABs are able to synthesise flavours. In the first step of the pathway, the amino acids are involved in dehydrogenation and transamination reactions with the formation of α-ketoacids, compounds which have a fundamental effect on flavour type and amount. Further, by decarboxylation reaction, α-ketoacids are transformed in aldehydes.

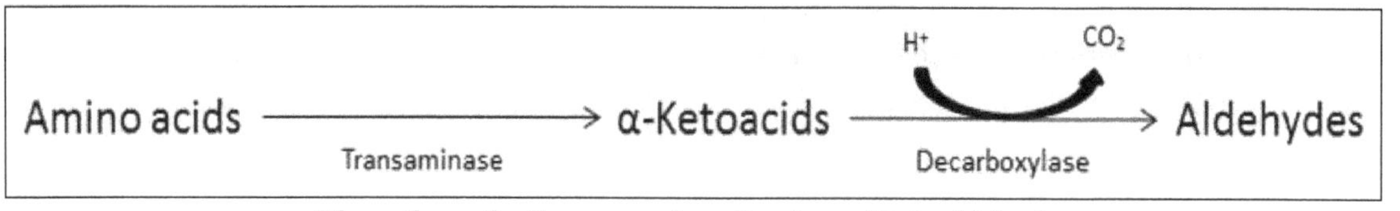

The pathway for the conversion of amino acids to aldehyde.

From decarboxylation reaction, a proton is consumed in the process, and the product is exported from the cell, resulting in an increase of the intracellular pH. Additionally, the aldehydes are transformed into alcohols or carboxylic acids by dehydrogenation, a majority of these compounds being flavour compounds. Several enzymes, for example, α-ketoacids, can thus be considered as intermediates involved in both biosynthesis and degradation of amino acids. Since branched-chain amino acids (Val, Ile, Leu), aromatic amino acids (Trp, Tyr, Phe) and sulphur-containing amino acids (Cys, Met) are important precursors of flavour compounds, the genome of Lactococcus lactis IL1403 was screened for gene-encoding enzymes of the biosynthetic pathways for these amino acids. At least 12 aminotransferases of the Escherichia coli are found to be encoded in the L. lactis IL1403 genome sequence. By knowing the enzyme and metabolic pathway, new potential flavours are expected to be biosynthesise for industrial applications. Theoretically, there are three pathways for the formation of α-ketoglutarate by bacteria using glutamate, citrate and pyruvate.

Amino Acids' First Specific Degradation Pathway

In the first step, the glutamate dehydrogenase pathway produces α-ketoglutarate directly from oxidative deamination of glutamate, utilising NAD^{+}, NADP or both as cofactor. NADP-dependent activity was detected in most Lactobacillus plantarum strains and in several Lactobacillus lactis, Lactobacillus paracasei and S. thermophiles strains, whereas NAD^{+}-dependent activity was observed in only a few L. lactis and S. thermophilus strains. Moreover, it has been demonstrated that the ability of LAB to produce aroma compounds from amino acids is closely related to their glutamate dehydrogenase activity. Conversion of amino acids to aroma compounds by LAB was limited by the lack of α-ketoacid acceptor for transamination reactions. Indeed, the addition of α-ketoglutarate to culture medium enhanced their aroma by increasing the amino acid catabolism.

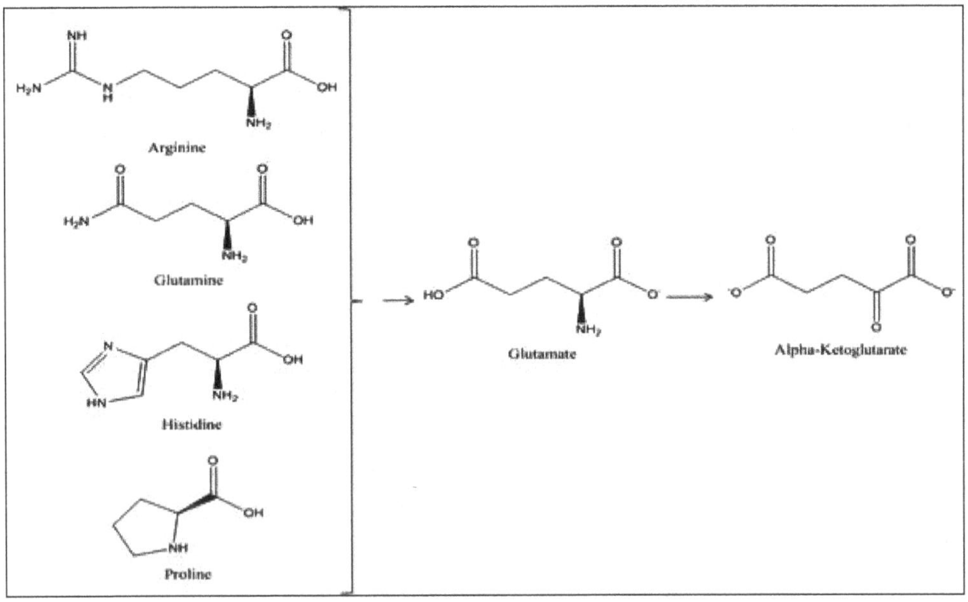

The glutamate dehydrogenase pathway of the amino acids.

α-Ketoglutarate is the best α-ketoacid acceptor for amino acid transamination by L. lactis. Another α-ketoacid that can also be used is pyruvate, but the aminotransferase activities were 40 times lower than with α-ketoglutarate. However, for some lactobacilli strains, pyruvate appeared to be an acceptor as efficient as α-ketoglutarate. A L. lactis strain genetically modified overexpresses a gene encoding a catabolic glutamate dehydrogenase, which catalyses the deamination of glutamate to α-ketoglutarate and, therefore, greatly increased the conversion of amino acids to potent aroma compounds. Pediococcus pentosaceus, Lactobacillus brevis, Lactobacillus curvatus and Lactobacillus fermentum inoculation leads to the conversion of glutamine to glutamic acid and NH_3.

Different amino acids have diverse amino peptidase, with characteristic activity on amino acids. Lb. fermentum IMDO 130101 possesses an arginine deiminase pathway which is modulated by environmental pH. This converts arginine into ornithine via citrulline while producing ammonia and ATP but at the same time has the ability to catabolise arginine to α-ketoglutarate by glutamate formation. The proline catabolism by Saccharomyces cerevisiae also leads to flavour biosynthesis, the intermediary compound being glutamate, which is further degraded to aroma compounds.

Amino Acids' Second Specific Degradation Pathway

The second possible pathway is the citrate-oxaloacetate pathway, which leads to α-ketoglutarate production from citrate and glutamate, by successive action of citrate permease, citrate lyase and aspartate aminotransferase (AspAT). Citrate permease allows citrate uptake inside the cells with the citrate catabolism initiation by transforming citrate to oxaloacetate. Oxaloacetate can then be transformed into aspartate and α-ketoglutarate, in the presence of glutamate, by an aspartate aminotransferase. For L. lactis species, only the diacetylactis subspecies possesses citrate permease and citrate lyase, but in this subspecies, oxaloacetate is mainly decarboxylated to pyruvate, which is then transformed to lactate, acetate, and diacetyl.

L. lactis IFPL326 strain showed the highest aminotransferase activity towards isoleucine, which is a specific substrate for the Lactococcus branched-chain aminotransferase. This LAB in combination with other strains which has α-ketoacid decarboxylase with high specificity for branched-chain degradation can be used for the obtaining of high yield of isoleucine-derived volatile compounds (2-methyl-1-butanol, 2-methylbutanal and phenylacetaldehyde) in the incubated milk. For example, 2-methyl-1-butanol is one of the components of the black truffle (Tuber melanosporum) aroma. Some Lactobacillus helveticus strains have been capable of diacetyl biosynthesis from α-aceto-α-hydroxybutyrate, an intermediate of isoleucine metabolism. Other research show that isoleucine catabolism leads to the formation of α-keto-β-methyl valerate.

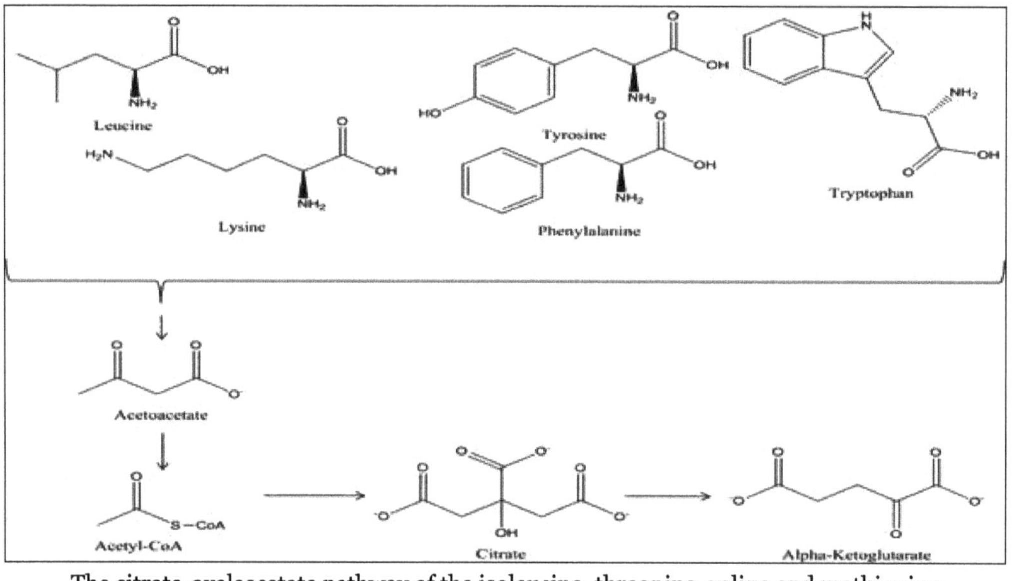

The citrate-oxaloacetate pathway of the isoleucine, threonine, valine and methionine (AspAT—aspartate aminotransferase).

On the other hand, the valine catabolism by non-oxidative enzymatic decarboxylation leads to the formation of α-ketoisovalerate in L. lactis fermentation that uses α-ketoisovalerate decarboxylase. The high specificity of the L. lactis α-ketoisovalerate decarboxylase permits to be a key controlling step in the formation of branched-chain aldehydes. Lactococcus strains combined with L. lactis IFPL730 for incubation in milk lead to production of aldehydes, without the necessity of exogenous α-ketoglutarate addition, and the production of different flavour compounds, like 2-methyl-1-propanol, 2-methylpropanal (straw fragrances) and 2-methyl-1-propionic acid (rum-like odour), was observed.

In another study, the L. lactis aromatic aminotransferase converts aromatic amino acids but also leucine and methionine. The methionine conversion was in lower concentration than isoleucine, leucine and valine. Aminotransferase activity requires α-ketoglutarate with the formation of 4-methylthio-2-ketobutyric acid which can be converted to methane-thiol, via a thiamine pyrophosphate-dependent decarboxylase that produces 3-methylthiopropanal, dimethyl sulphide and dimethyl disulphide. It is important to mention that methional is a notable flavour used in potato-based snacks, while dimethyl disulphide has a garlic-like aroma. During cheese ripening, cystathionine β-lyase can convert methionine to various volatile flavour compounds, but in bacteria its physiological function is the conversion of cystathionine to homocysteine, which is the penultimate step of methionine biosynthesis. In other researches, beside aroma abovementioned, obtained from methionine catabolism, phenylacetaldehyde (with honey-like aroma) was identified by Lb. plantarum UC1001, S. thermophilus and Lb. helveticus biosynthesis.

The biosynthesis of the diacetyl from aspartate by some Lactobacillus strains has been reported. Aspartic acid under the aminotransferase action may generate acetoin and diacetyl by Lactobacillus casei GCRL163 and Lactobacillus strains. Three Lb. paracasei subsp. paracasei strains (CHCC 2115, 4256 and 5583) had different expression of aspartate aminotransferase activities against aspartate. Another study shows that all the LAB strains with citrate metabolism can biosynthesise diacetyl and acetoin by aspartate metabolism with the formation of the unstable α-acetolactate.

LAB protein degradation determines the formation of free amino acids that vary in their concentration over time. Leucine has been reported to be dominant amino acid in Cheddar cheese after 6 months of maturation. Leucine catabolism leads to the formation of α-keto-isocaproate and generates aroma like 3-methylbutanal (cheesy, chocolate, malt), 3-methylbutanoic acid (cheesy, sweaty), phenylacetaldehyde and 2-hydroxy-4-methyl pentanoic acid methyl ester. On the other hand, under Lb. plantarum UC1001, S. thermophilus and Lb. helveticus catabolism of lysine results in hexanoic acid, with a cheesy aroma.

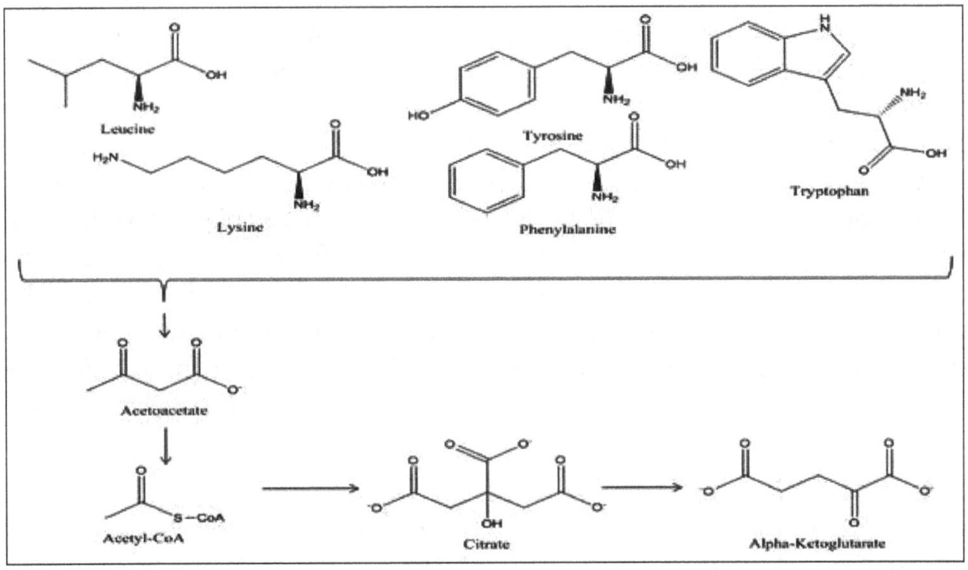

The citrate-oxaloacetate pathway of the leucine, lysine, tyrosine, phenylalanine and tryptophan.

L. lactis subsp. diacetylactis and Lactococcus lactis subsp. cremoris strains used in the cheese manufacturing are able to degrade phenylalanine and leucine in the presence of citrate and glutamate. This is possible due to the fact that this strains use α-ketoacids (pyruvate and α-ketoglutarate) as acceptor for transamination reaction, produced from citrate metabolism. To balance the α-ketoglutarate biosynthesis (because this is the best acceptor for L. lactis aminotransferase and the pyruvate is an enzyme used in many pathways), a selection of a strain with a high aspartate aminotransferase activity and low oxaloacetate decarboxylase activity may be introduced into co-fermentation. From the phenylalanine catabolism resulted in phenylacetaldehyde, a floral aroma and a key odour compound in hard and semihard cheese varieties. This aroma in combination with p-cresol, phenyl-ethanol, indole and skatole can result in undesirable odour that contributes to putrid, faecal or unclean flavours in cheese. By using a specific strain, undesirable flavours can be avoided. L. lactis degrades 49% of initial phenylalanine with the biosynthesis of phenyl-lactate, phenyl-acetate, benzaldehyde (which has an almond-like odour) and phenyl-ethanol (with a floral odour) and 22% of initial leucine in milk fermentation with the formation of the hydroxyl-isocaproate and isovalerate (menthol aroma). Lb. plantarum UC1001, S. thermophilus and Lb. helveticus can produce 2-phenethyl alcohol (rose-like aroma) and phenylacetaldehyde (floral fragrances) along with flavours named from phenylalanine catabolism. These three strains can also produce benzaldehyde from tryptophan catabolism. Tyrosine is degraded by Brevibacterium linens 47 by phenylalanine pathway.

Amino Acids' Third Specific Degradation Pathway

The third pathway is the citrate-isocitrate pathway which utilises the oxidative branch of the tricarboxylic acid cycle leading to the production of α-ketoglutarate from either pyruvate or citrate with the action of pyruvate dehydrogenase, pyruvate carboxylase, citrate synthase, aconitase and isocitrate dehydrogenase. Pyruvate dehydrogenase and pyruvate carboxylase are necessary to degrade pyruvate into acetyl-CoA and oxaloacetate, respectively, both used by citrate synthase to synthesise citrate. Citrate is then transformed by aconitase into isocitrate, which is finally oxidised to α-ketoglutarate by isocitrate dehydrogenase.

Pediococcus acidilactici and P. pentosaceus can convert alanine to pyruvate by α-ketoacid intermediary pyruvate which further is converted to flavour compounds. The diacetyl and acetoin are produced via citrate metabolism by citrate-positive LAB (L. lactis, Lb. casei) through aspartate catabolism described by the l-aspartate-l-alanine-pyruvate steps. The degradation pathway of the alanine by Lb. plantarum UC1001, S. thermophilus and Lb. helveticus leads to the production of acetic acid and ethanol, while from glycine catabolism resulted in acetic acid. Other microorganisms degrade glycine to pyruvate with the formation of serine as intermediary compound, and then the pyruvate is used as a precursor to flavour biosynthesis. By oxidative deamination of the serine under the Lb. plantarum UC1001 metabolism, acetic acid is detected, while P. pentosaceus and P. acidilactici are able to produce diacetyl from pyruvate and l-serine.

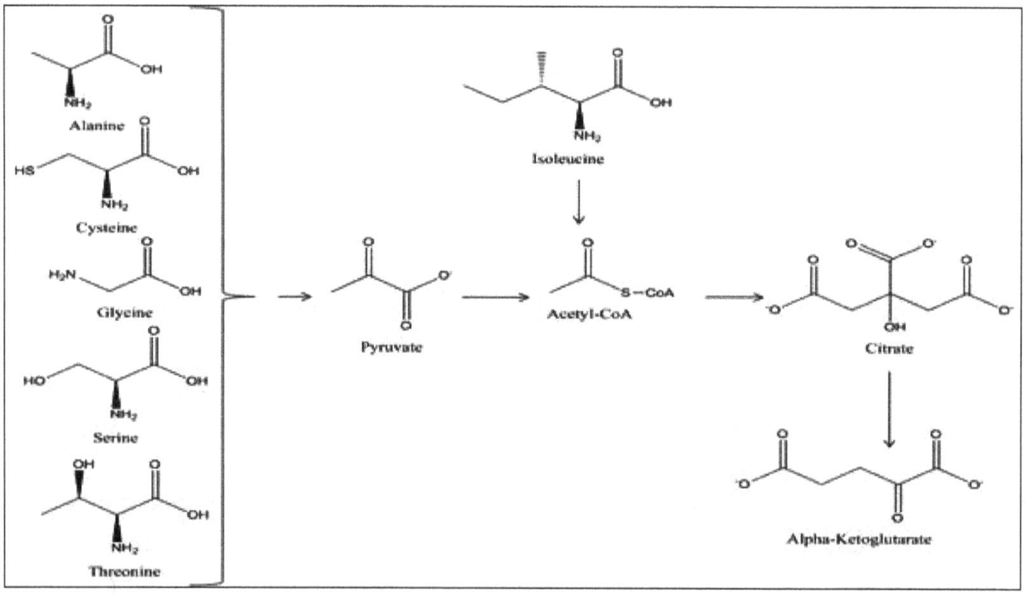

The citrate-isocitrate pathway of the alanine, cysteine, glycine, serine and threonine.

Cysteine is catabolised by α-ketoacid enzyme with synthesis of 3-mercaptopyruvate, which by elimination of hydrogen sulphide, lead to the obtaining of pyruvate, used as a precursor for flavour biosynthesis.

Lb. helveticus and S. thermophilus can produce acetaldehyde from threonine by the breakdown of the amino acid with threonine aldolase into glycine and acetaldehyde. Acetaldehyde levels increase together with threonine levels, in cheeses during ripening. Branched aldehydes are produced from the catabolism of branched amino acids, but they do not accumulate in cheese because they are quickly converted into the corresponding alcohols. Lb. plantarum UC1001, S. thermophilus and Lb. helveticus can produce propionic acid from threonine catabolism.

The Influence of Carbon Sources

The carbon source is very important for the microbial growth because it is the principal resource for energy production. In the same time, for aroma biosynthesis sugars with a low-molecular weight are requested to be used, which are at the same time a flavour precursor.

The addition of sucrose in the culture media stimulates the flavour biosynthesis for yeasts and LABs. Di Cagno and co-workers supplemented with sucrose the tomato juice that is subjected to LAB fermentation in order to stimulate the flavour biosynthesis and to reduce the intrinsic flavour acidity of tomatoes. By inoculation of six prebiotic strains in the milk culture media, supplemented with 0.75% fructose (w/v), a desired aroma for the final product was obtained.

De Figueroa and co-workers demonstrated that Lactobacillus rhamnosus ATCC 7469 can use lactate as the sole energy source and, at the same time, is able to grow with citrate as sole energy source and to produce diacetyl and acetoin. The enzyme activity of this strain is increasing with the increase of temperature from 22 to 37 °C. Therefore, the presence of a high pyruvate amount lied to a high production of acetolactate, diacetyl and acetoin. At the same time, when the glucose level is high, diacetyl and acetoin in low concentration are produced by Lb. rhamnosus ATCC 7469. Some LAB species like Lb. rhamnosus and Lb. plantarum are able to grow on citrate as a single carbon source and consequently to produce diacetyl.

In sourdoughs, for example, flavour compounds are produced by LAB and yeasts individually or by their interactions. S. cerevisiae produced more volatile compounds than Candida guilliermondii, but quantity of volatile flavour compounds can be improved by the addition of glucose, of sucrose and less of maltose. Addition of fructose, glucose or maltose to the dough increases LAB contributions to volatile formation in baking. Escamilla-Hurtado and co-workers prepared a semisolid maize-based culture media and grew a mixed cultures formed by P. pentosaceus MITJ-10 and Lactobacillus acidophilus Hansen 1748 obtaining 779.56 mg/kg diacetyl after 12 h of exponential growth. Enterococcus faecium FAIR-E 198 can grow on xylose, glucose and lactose and converted by biosynthesis the citrate in diacetyl. However, in non-fermentable conditions, the acetoin yield is decreased in the strain fermentation. Also, other species like Leuconostoc can use xylose as a sole energy source for diacetyl biosynthesis. Lb. casei GCRL163 strain was studied in a medium supplemented with different concentrations of lactose, but the maximum growth was registered for only 1% lactose in medium with no significant aroma biosynthesis.

The Influence of the Mineral Composition of the Medium

The minerals are very important in culture media of the microorganism because they are used as a cofactor in enzymatic activity. All enzymes have a metal as a coordinative element, and the enzyme activity depends on it. At the same time, the salt concentration is very important because it dictates the osmotic pressure and flavour improvements. Similarly, aldehydes can also be generated by chemical oxidation of α-ketoacids catalysed by bivalent cations. Manganese and magnesium sulphate enhanced both biomass and aroma development of 52 different yeasts by obtaining 96.05 mg/L acetaldehyde for Candida lipolytica and 3.58 mg/L diacetyl for Candida globosa.

Recently, two manganese transport systems of Lb. plantarum have been characterised. These systems, which are implicated in mineral uptake, convert phenylalanine to benzaldehyde by initiation of a pyridoxal 50-phosphate-dependent aminotransferase. The phenyl-pyruvic acid is obtained after conversion, which is further chemically transformed to benzaldehyde in the presence of oxygen and manganese.

The Influence of Temperature

The flavour biosynthesis by microorganism is strongly influenced by the temperature of fermentation. De Figueroa and co-workers demonstrate that Lb. rhamnosus ATCC 7469 produce diacetyl and acetoin from citrate within a temperature interval of 22–45 °C. The biosynthesis amount of diacetyl increased in the temperature interval between 30 and 37 °C with maximum production at 48 h. For the fermentations made at different temperatures, as 22 and 45 °C, the maximum aroma biosynthesis was reached at 24 h of incubation, and the level of the acetoin and diacetyl was 4.1 time higher at 37 °C than at 22 °C. Moreover, the highest efficiency of the conversion of citrate into diacetyl and acetoin was obtained at 37 °C. At the same time, citrate transport and incorporation in microbial system reach maximum at the 37 °C. Another effect of the temperature is on the enzymatic systems. The activities of citrate lyase and NADH oxidase reach a maximum at 37 °C when the temperature is increased from 22 to 45 °C.

On the other hand, lower incubation temperature tends to selectively promote growth rate of the Leuconostoc and L. lactis ssp. cremoris strains, while the higher temperatures will favour Lb. rhamnosus and L. lactis ssp. lactis strains. The inoculation concentration has also a significant influence on the aroma production. The acetaldehyde biosynthesis by Leuconostoc or Lactobacillus is not influenced by the temperature changes.

For the yeast fermentations (S. cerevisiae), the increased temperature from 24 to 30 °C leads to increasing of the acetaldehyde biosynthesis. Besides yeast, acetic acid bacteria can biosynthesise acetaldehyde at concentrations up to 250 mg/L. For this fermentations type, acetaldehyde tends to accumulate under low oxygen level and ethanol concentration higher than 10%.

The Influence of Aeration

The presence of oxygen strongly influences the microbial growth and the flavour biosynthesis. The aerobic microorganism's metabolism is oxygen dependent and is mandatory for flavour biosynthesis pathway. For example, Lb. casei grown under aeration conditions leads to higher diacetyl amount biosynthesise in Cheddar cheese than an anaerobic starter culture. Another example is for E. faecium FAIR-E 198 strain growth which biosynthesise diacetyl only in aerobic conditions.

Acetaldehyde Biosynthesis Pathway

Acetaldehyde besides being a major component of tobacco smoke is the primary metabolite of ethanol. Commercially, it is obtained by Wacker process of ethylene oxidation in strong acid solutions using as catalysts $PdCl_2$-$CuCl_2$ of crude oil, but this method is not very sustainable. The trend demands are for obtained acetaldehyde from renewable raw materials like sugars from biomass or synthesise from lactic acid. Due to its high reactivity derived from containing two conjugated hydroxyls and one carboxylic group, lactic acid (LA) is an attractive feedstock for chemical production, being in torn biosynthesise at low costs by glucose and xylose fermentation. The acetaldehyde may be produced by decarbonylation or decarboxylation of LA in the presence of aluminium phosphates and magnesium aluminate spinels, reaction promoted by acid catalysts.

Acetaldehyde represents a secondary metabolite in alcoholic fermentation of yeasts, being a precursor of the ethanol production in beer and wine. It is the most important carbonyl compound produced during alcoholic fermentation in concentrations between 10 and 200 mg/L depending on technological factors, such as culture medium composition, pH, fermentation temperature, aeration and SO_2 concentration and on the yeast strain used. Acetaldehyde is biosynthesized from glucose by the glycolytic pathway enzyme pyruvate decarboxylase. At the beginning, two molecules of pyruvate resulted from glucose glycolysis, and by pyruvate decarboxylation, the secondary acetyl-CoA product is obtained. Furthermore, two acetaldehyde molecules are resulted under alcohol dehydrogenase action on the acetyl-CoA compound. The high peak value of acetaldehyde biosynthesis is reached during the early fermentation phases, being then partly re-catabolised by yeast, or is combined with polyphenols or other compounds in the wine being a very reactive compound.

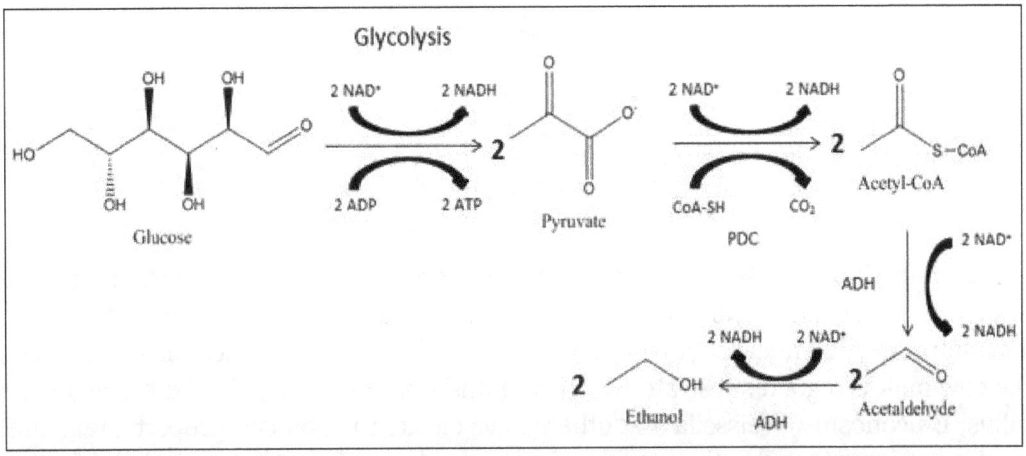

Acetaldehyde biosynthesis pathway. PDC, pyruvate decarboxylase; ADH, alcohol dehydrogenase.

Another species that can produce acetaldehyde is the acetic acid bacteria (AAB) characteristic from grape microorganism equipment. This class of microorganism has another biosynthesis pathway, oxidising ethanol to acetaldehyde and acetic acid in concentrations up to 250 mg/L. At ethanol concentrations higher than 10% (v/v) and under low oxygen conditions, acetaldehyde tends to accumulate, in other conditions being oxidised to acetic acid. The yeasts reported to biosynthesize acetaldehyde in high amounts are S. cerevisiae with 0.5–286 mg/L and Kloeckera apiculata with 9.5–66 mg/L.

Taken into account that lactic acid bacteria are also responsible for acetaldehyde biosynthesis, being a typical flavour component of yoghurt responsible for pungent and fruity flavour, this microorganism has attracted the researcher attention. By using recombinant microbial processes for biotransformation approach, Balagurunathan and co-workers engineered an E. coli strain for acetaldehyde production from glucose. They introduced the pyruvate decarboxylase from Zymomonas mobilis and NADH oxidase from L. lactis in the E. coli strain genome, and the results confirmed that around 37% of the glucose consumed could be redirected towards acetaldehyde biosynthesis under anaerobic conditions. The mass yield of acetaldehyde obtained is 0.18 g/g glucose, this being the highest mass reported for microbial acetaldehyde production. The main disadvantage of biosynthesise acetaldehyde is the high toxicity on the microbial cells.

Diacetyl Biosynthesis Pathway

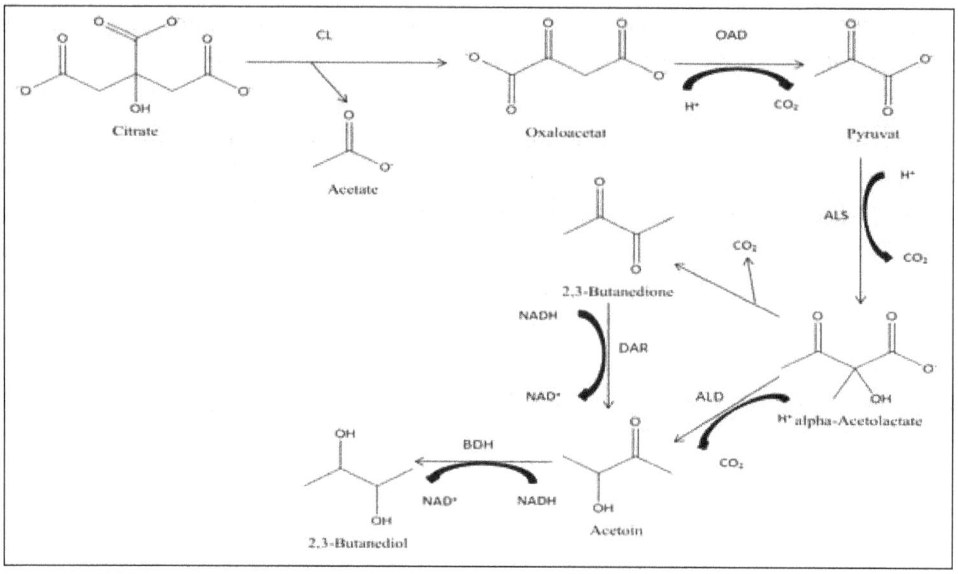

Diacetyl biosynthesis pathway. CL, citrate lyase; OAD, oxaloacetate decarboxylase; ALS, α-acetolactate synthase; ALD, α-acetolactate decarboxylase; DAR, acetoin/diacetyl reductase; BDH, butanediol dehydrogenase.

Diacetyl (2,3 butanedione) is the typical butter flavour/aroma, commonly found in fermented dairy products, such as butter, sour cream and yoghurt, and important for cheese aroma. Moreover, L. lactis is a safe flavour production microorganism. The precursor for diacetyl and acetoin biosynthesis is citric acid, while cow milk is a good substrate which contains 1750 mg citrate/L. Genera as Enterococcus, Lactobacillus, Leuconostoc, Weissella and others have citrate metabolism property, regulated by different gene expression adapted to the specific microorganism. Different transcriptional factors belonging to DeoR and GntR family mediate transcriptional activation in the presence of the substrate.

Citrate metabolism is the principal generator of sensory characteristics of the milk final products. One of the LAB used globally as starter fermentation is L. lactis biovar diacetylactis where citrate and acetoin/diacetyl pathway increases the intracellular level of pyruvate and is coordinated and expressed at low pH.

For LAB, citrate fermentation is involved in cheese flavour and quality. In the first step of citrate metabolism catalysed by citrate lyase, oxaloacetate is obtained. The oxaloacetate is decarboxylated via oxaloacetate decarboxylase generating pyruvate. In L. lactis, genes associated with citrate metabolism are organised in two operons. One operon, citQRP, is involved in citrate transport, and the other operon, citM-citI-citCDEFXG, which is encoded for citrate lyase, is involved in citrate conversion to pyruvate. In this step, the proton motive force is generated (PMF). In LAB enzymatic equipment, two types of oxaloacetate decarboxylase are presented. One type is a soluble oxaloacetate decarboxylase belonging to the malic enzyme family. This enzyme is present in L. lactis and Weissella mesenteroides and other LAB strains, catalysing the conversion of oxaloacetate from citrate to pyruvate in the presence of divalent metals. The second type is an oxaloacetate decarboxylase membrane complex, which is a biotin-dependent decarboxylase. This consists of four polypeptides and is found in Enterococcus faecalis and Lb. casei.

The pyruvate obtained is condensed by α-acetolactate synthase with the formation of α-acetolactate which is chemically unstable. The α-acetolactate formed by the action of α-acetolactate decarboxylase can be converted to acetoin or diacetyl in a non-enzymatic oxidative decarboxylation reaction, the biosynthesis being more evident at pH 4.5. This pathway could be also completed with additional enzymes such as acetoin/diacetyl reductase and butanediol dehydrogenase.

There have been many attempts to redirect the metabolism of various microorganisms for improving diacetyl formation, by classical mutagenesis or directed genetic engineering trying the improvement of by-product formation, but the results were not very promising.

Flavour Determination and Quantification

The flavour determination was made by different instrumental analytical methods. The common one is the solid-phase micro-extraction analysis, in which different substrates are used for solid phase.

Table: The solid-phase micro-extraction conditions for flavour analysis.

Solid-phase material	GC column
Carboxen/polydimethylsiloxane, 85 μm film thickness, 220 °C work temperature.	Zebron ZB-624, D-0.25 mm; 1.4 μm film thickness; composition, 94% dimethyl polysiloxane; 6% cyano-propyl-phenyl; 60 m long.
Polydimethylsiloxane with 10% embedded activated carbon phase (PDMS/AC), 50 μm film thickness, 250 °C work temperature.	HP-5MS capillary column, 5% phenyl methyl silicone, 320 μm × 1.0 μm, 60 m long.
Polyacrylate bonded to silica core, 85 μm film thickness, 220 °C work temperature.	HP-INNO-WAX polyethylene glycol capillary column, 250 μm × 0.5 μm, 60 m long.
Polydimethylsiloxane fibre, 250 °C work temperature.	DB5 capillary column, 0.32 μm internal diameter, 1 μm film thickness, 60 m long.
StableFlex divinylbenzene/carboxen/polydimethylsiloxane (DVB/CAR/PDMS) coated fibre, 250 °C work temperature.	ZB-WAXplus polyethylene glycol capillary column, 0.25 mm internal diameter; 0.50 μm film thickness, 60 m long.
Silica fibre covered by Carboxen Polydimethylsiloxane (CAR-PDMS), 75 μm film thickness, 250 °C work temperature.	CP-Wax 52 CB polyethylene glycol coated, 0.32 mm, 1.2 mm film thickness, 50 m long.
Polydimethylsiloxane divinylbenzene (PDMS_DVB) SPME fibre, 250 °C work temperature.	SUPELCOWAX™ 10 capillary column, 0.1 mm, 0.1 μm film thickness, 10 m long.

Other analysis techniques of flavours are (i) By purge and trap method and (ii) GC-MS separation and identification or (iii) By proton transfer reaction mass spectrometry using PTR-MS. In some cases, the flavour determination was made enzymatically or by derivatisation with dinitrophenylhydrazine (DNPH)-acetonitrile reagent, and then the compounds are analysed by HPLC analysis with detection at 360 nm.

Applications of Natural Flavours

Flavour release from food during consumption in the mouth is important in flavour perception and influenced by food matrix. Since food matrix changes biochemically and physically during eating, the food flavour microencapsulation results in controlled release for specific situations. Different natural and synthetic polymers were used for microcapsule fabrication, of which alginate-whey protein compounds have been found to be suitable as vehicle for diacetyl flavour delivery.

Nowadays, the delivering of antimicrobial volatiles from polymeric systems, in a controlled manner, gained an increasing interest. In food industry, diacetyl is used not only as an approved food additive but also for food preservation, due to its antimicrobial activities. Diacetyl has been shown to be bactericidal against E. coli and Staphylococcus aureus at a concentration as low as 100 ppm. The effects of diacetyl on the quality of ground beef were evaluated when diacetyl was used in modified-atmosphere packaging in conjunction with 20% CO_2. A delayed spoilage of ground beef and the maintenance of the fresh colour and odour were observed for this product. The inhibitory effects of diacetyl combining with reuterin, against E. coli, Salmonella enteritidis and Listeria monocytogenes in milk, suggested that these LAB metabolites are potential for pathogen control in dairy products.

Strains of Lactobacillus and bifidobacteria could produce diacetyl in concentrations up to 30 mg/mL suggesting its potential to exhibit dermal antimicrobial activities, with greater sensitivity against Gram-negative bacteria (such as Pseudomonas aeruginosa, Pasteurella multocida, Borrelia burgdorferi, Salmonella typhi, Bartonella sp., Klebsiella rhinoscleromatis, Vibrio vulnificus and Helicobacter pylori) and fungi as compared to Gram-positive bacteria.

Another direction of diacetyl utilisation is related to active packaging systems. Thus, the controlled release of different volatile antimicrobial compounds was tested for packaging obtained from two or more poly(ethylene glycol) polymers of different molecular weights and/or a mixture of poly(-lactic acid) and poly(ethylene oxide).

Sweeteners

Sweeteners are food additives that are used to improve the taste of several foods. Many of the natural sweeteners contribute to energy value. Calorie conscious consumers want to enjoy food without putting on weight. Some sweeteners have been modified to produce some valued sweeteners e.g. polyols that are gaining prominence these days. Artificial sweeteners, on the other hand, are compounds that have very little or no nutritional value.

Sugar Alcohols/Polyols

Mannitol, sorbitol, and maltitol occur naturally in fruits. Xylitol is a normal intermediate product in the metabolism of carbohydrates in fruits and vegetables. These are alcohol derivatives of sugar and provide fewer calories than sucrose. Polyols are stable up to 160 °C. It is popularly used as a bulking agent in the food industry in order to improve the taste, mouth-feel and texture of the food. Sugar alcohols add bulk and texture to food such as chewing gum and hard candies.

Polyols are absorbed by passive diffusion and metabolized slowly. They produce a low glycemic response. Intake of large amounts of polyols exerts a laxative effect causing abdominal cramps and diarrhea. Polyols also prevent dental caries when used a substitute for sugar in different products.

Sugar alcohols are found in diabetic candies, chewing gums, and as coating on gums. Products containing sugar alcohols may carry the label 'sugarless', 'sugar-free', or 'no-sugar', but they are still carbohydrates and supply calories.

Sorbitol

General characteristics: It is a hexahydric alcohol, obtained by catalytic hydrogenation of D-glucose or invert sugar at high temperatures and pressure, followed by ion-exchange treatment. It is a white, odourless, hygroscopic crystalline powder, easily water soluble and possesses a sweet cool and pleasant taste. It is a nutritive sweetener providing 2.6 k.cal./g.

Advantages for inclusion in food products:

- It is non-cariogenic, non-reducing and very resistant to change by heat.
- It acts as a flavor carrier, retards sugar crystallization, controls freezing properties and contributes to the viscosity and texture of frozen desserts.
- It is useful in formulating 'Soft ice creams' and 'Diabetic mix'.
- When employed in 'Diabetic product', it replaces sugar in providing some of the bulk and mouth feel, and also some sweetness.

Disadvantages for use in food products:

- The relative cost/unit sweetness of sorbitol is higher than that of sucrose.
- Indiscriminate use has been associated with a laxative effect; laxative threshold being 71.9 ± 4.9 g/day.

Lactitol

General characteristics: It is a disaccharide comprising of galactose and sorbitol, produced by the reduction of glucose moiety of lactose from e.g. whey. It is a non-hydroscopic, slightly less soluble than sucrose and is miscible with other sugars and sugar alcohols. The effect on freezing point (FP) is almost the same as that exerted by sucrose. The cooling effect is slightly higher than that of sucrose. Lactitol is being sold under trade name 'Lacty' by CCA BioChem, Netherlands.

Advantages for use in food products:

- It can reduce the risk of dental caries.
- It can be used in formulating 'diabetic ice cream'.
- Since it provides low energy value (i.e. 2.03 k.cal./g), it is used to formulate reduced-calorie frozen dessert.

- It provides a creamy flavor and a bland, sweet taste with no aftertaste.
- It is preferred over sorbitol due to its cleaner sweetness.
- It provides the same bulk and texture as given by sucrose.
- It does not undergo browning reactions, due to absence of free carbonyl group.

Disadvantages for use in food products:

- It is relatively expensive i.e. about $1/3^{rd}$ more expensive than sorbitol.
- It has a slight laxative effect, hence a daily maximum intake of 40 g has been recommended.

Xylitol

General characteristics: It is a pentahydric alcohol naturally occurring in fruits and vegetables. Commercially, it is obtained by acid hydrolysis of xylan, a polysaccharide component of hemicelluloses in plants to xylose. Xylose is purified, silated and hydrogenated into xylitol. Hydrogenation of xylose converts the sugar (aldehyde) into a primary alcohol. It is a nutritive sweetener providing 2.4 k.cal./g and is readily soluble in water.

Advantages for inclusion in food products:

- It has an excellent taste and has a cooling effect greater than sorbitol.
- It is reported to cause 90% reduction in dental caries and is claimed to possess 'tooth-rebuilding' properties.
- It can be used as an excellent sugar substitute for diabetics.
- It is highly suited for manufacture of 'soft ice creams', since it produces product that is softer in consistency.

Disadvantages for use in food products:

- It has a laxative effect at high dosage; the tolerable limit being 50 g/day.
- It is costly compared to sucrose.
- It is suspected of being carcinogenic and long term intake is implicated with derangement of liver metabolism.

Maltitol

Maltitol is about 90% as sweet as sugar, non-cariogenic, and has half the calories of sugar. Maltitol is useful in the production of sweets, including sugarless hard candies, chewing gum, chocolates, baked goods and ice cream.

Mannitol

Mannitol is about 50% as sweet as sugar. It is found naturally in mushrooms and trees. It is manufactured for use in food. It is often used as a dusting powder for chewing gum products.

Erythritol

Erythritol occurs naturally in fruits such as pears, melons and grapes, as well as foods such as mushrooms and fermentation-derived foods such as wine, soy sauce and cheese. Since 1990, Erythritol has been commercially produced and added to foods and beverages to provide sweetness, as well as enhance their taste and texture. Erythritol has the best tolerance level among all polyols. Since they are metabolized by the body more slowly than sucrose, they are useful in foods for people following special diets, such as a diabetic diet.

Isomalt

Isomalt is a substitute sugar made from the sugar sucrose. Isomalt is unique among replacement sugars because it retains almost all of the physical properties of real sugar. Isomalt absorbs little water, so products do not become sticky which is an added benefit for sugar artists, cake decorators and pastry chefs.

Non-Nutritive Sweeteners/Sugar Substitutes

Sugar substitutes are food additives that duplicate the taste of sugar in food, but do not supply food energy or calories. Some sugar substitutes are natural and others are synthetically produced. These synthetically produced sweeteners are generally called 'artificial sweeteners'.

These sweeteners are known as high-intensity sweeteners as they are required in minute quantities to provide the same sweetness as sugar. Examples of such food additives are saccharin, aspartame, sucralose, acesulfame-K, stevioside, cyclamate, etc. A person with diabetes would not like to miss this sweet taste and cut out 25% of his eating pleasure.

Saccharin

It is one the oldest non-nutritive sweeteners. It is a sulfonamide (benzoic sulfimide) produced from a compound found in coal tar. It is sold as a white powder for use as a table-top sweetener. Saccharin is available as salts of calcium and sodium. It is non-cariogenic and may help inhibit development of dental caries. The bitter, metallic aftertaste of saccharin can be eliminated by blending it with other sweeteners, especially aspartame. It is about 300 times sweeter than sucrose. Its melting point is 228 °C. It is used in a variety of foods such as beverages, jams, and baked goods. In India, it is widely used in food and beverage industry.

Saccharin was banned in 1977 as it was incriminated in the causation of bladder cancer in rats. This ban was revoked in 1991. Till late, a warning label was required for products containing saccharin. It has been recently removed from the carcinogen list. It may cause urticaria and eczema in some people. Intake of saccharin is restricted in pregnant women as it is avidly concentrated in the placenta.

Advantages for inclusion in food products:

- It has a Generally Recognized as Safe (GRAS) status.
- It is stable under a wide range of pH and temperature and over prolonged storage.
- The sweetener is not metabolized and thus non-caloric.

Disadvantages for use in food products:

- The sweetener exhibits a slow onset of sweetness, which increases to a maximum and persists.
- It has a bitter, metallic aftertaste and is unstable at high temperatures.

Aspartame

Aspartame is a high intensity sweetener, dipeptide in nature. It is made up of two amino acids, aspartic acid and methyl ester of phenylalanine. It is available as a white, colourless, crystalline powder which is slightly soluble in water, with maximum solubility at pH 2.2. It is unstable at high temperatures and is added in the end-cycle of cooking. Commercially aspartame is available as Equal, Nutrasweet, Saneeta and Spoonful. Aspartame has a flavor similar to sucrose, and also functions as a taste intensifier and enhancer. Aspartame is about 180-200 times sweeter than sucrose. It has a clean, sweet taste, no after taste, enhances and extends flavours, especially fruit flavours. Aspartame tends to denature or lose its sweet taste when heated.

It acts synergistically with other sweeteners. It is preferable to use this sweetener (@ 0.05-0.1%) in conjunction with either polydextrose (@ 11-12%) or low DE Corn syrup solids (CSS) (@ 15%) in order to obtain satisfactory quality frozen dessert.

Food and Drug Administration (FDA) has set an Acceptable Daily Intake (ADI) of 50 mg/kg body weight for aspartame. Aspartame was approved by FDA in 1981 for use in a variety of products, as a table-top sweetener and in carbonated beverages. In July 1993, approval was made for use in hard and soft candies, baked goods and mixes, non-alcoholic beverages and malt beverages. Frozen yoghurt is an ideal medium for aspartame sweetener. The slight acidity of yoghurt (pH ~ 4.5) results in a close to optimum pH for the stability of sweetener.

Unlike the alternate sweeteners saccharin, acesulfame K, neotame and sucralose, which have no nutritional value, aspartame contributes calories. Aspartame has been approved for use in yoghurt type products, dairy and non-dairy frozen desserts in UK, USA, etc. Food products that contain aspartame must carry a warning to people with 'phenylketonuria', a rare genetic disorder that prevents proper metabolism of phenylalanine.

The specific by-products of aspartame i.e. methanol, formaldehyde and phenylalanine have been studied. They are said to cause an array of side-effects, including allergic reactions in some people. It has also been linked to neuropsychiatric disorders like migraine headaches in sensitive individuals, panic attacks, mood changes, epileptic seizures.

Advantages for inclusion in food products:

- The absence of bitter character and sucrose-like flvaour (i.e. clean, sweet taste) makes this sweetener a source superior to other frequently used high intensity sweeteners like saccharin.
- It can be used in both HTST and UHT pasteurization systems with minimal loss, but it cannot withstand baking.
- It is non-cariogenic and helps in production of 'sugar-free' food items.

- Although it has an energy value of 4.0 k.cal./g, because of its low rate of usage, it is effective in energy reduction.
- It shows synergistic effect with fruit flavours and hence successfully used in orange sherbet to supply a maximum of 37.5% of the sweetness.
- It acts synergistically with other sweeteners.

Disadvantages for inclusion in food products:

- It can react with reducing sugar i.e. glucose, resulting in some loss of sweetness.
- It is costly compared to other intense sweeteners; however, the price is expected to decrease with increased usage.

Acesulfame-K

It is a potassium salt of dimethyl-oxathiozine dioxide in nature; a cyclic sulfonamide. It is a colourless, odourless, crystalline product, readily soluble in water and the taste is quickly perceptible without lingering. Acesulfame-K has a grittier texture than direct sugar but has a slight bitter aftertaste. Acesulfame-K is similar to aspartame in sweetening power, but holds up better to heat and costs less. Marketed as Sunette, Sweet One and Diabeti-sweet, the sweetener was approved by FDA in 1988 for limited use in products such as chewing gums and dry beverage mixes. In 1998, the FDA approved acesulfame-K for use in soft drinks. Other applications include instant coffee and tea, puddings, gelatins, chewing gum, dairy product analogues, salad dressings, confectionery, baked products and as tabletop sweeteners. Abroad, it is being used increasingly along with aspartame in many foods and beverages and is second in popularity to aspartame.

Acesulfame-K, a derivative of acetoacetic acid, is not metabolized and is excreted from the body unchanged. Caution is required in individuals on potassium restricted diet or having sulfa antibiotic based allergy.

Advantages for use in food products:

- It is a non-caloric sweetener.
- It can withstand high temperature cooking and baking and is stable.
- It has synergistic sweetness effect with aspartame, cyclamate and saccharin.
- The sweetness intensity being slightly higher in acidic condition (> 3.0 ph), the sweetener would be of particular advantage in sherbets, ices and frozen yoghurt.
- During prolonged storage.

Sucralose

Sucralose is a modification of the sucrose molecule i.e. a chlorinated version known as tri-chlorogalactosucrose. It is produced from sucrose by chlorination of 3-hydroxyl groups. Sucralose retains its sweetness over wide range of temperatures. It is marketed under brand name 'Splenda' abroad. In India, it is available under brand name 'Zero' and 'Sugarite'.

Approved by the FDA in 1998, Sucralose is the only alternative sweetener made from sugar. It is 600 times sweetener than sucrose (1.6 calories per tea spoon). Sucralose is derived from sugar through a patented, multi-step process that creates an extremely stable substance unable to be absorbed by the human body. Sucralose is not absorbed and gets excreted unchanged in the body. It does not cause any rise in blood glucose levels. It is excreted in the urine virtually unchanged. Sucralose does not contribute to tooth decay. It may cause shrinkage of the thymus gland, enlargement of the liver and kidney.

Owing to its stability, it is used in a range of products such as carbonated soft drinks, low-calorie fruit drinks, maple syrup, canned fruit, low-calorie fruit drinks, confections, baked goods, sauces, and syrups. Sucralose can be used as a sweetener in nutritional supplements, medical foods and vitamin/mineral supplements. Sucralose can be found as a table top sweetener too.

Advantages for its use in food products:

- It offers unique combination of sweet, sugar-like taste, free of any unpleasant aftertaste.
- It has thermal stability and hence not adversely affected during processing of mixes.
- It is non-cariogenic as well as non-caloric sweetener; it helps in calorie reduction.
- It is resistant to hydrolysis and breakdown by microorganisms.

Glycyrrhizin

It is the root extract of an Asian and European shrub, commonly known as licorice. It is a pentacyclic triterpene saponin glycoside. It is mainly used as coating for candies and chewing tobacco. Ammonium glycyrrhizinate is commercially available. Ammonium glycyrrhizinate and glycyrrhizin are 100 and 50 times sweeter than sucrose respectively.

Ingestion of large amounts of food containing glycyrrhizin may cause headache, lethargy, sodium and water retention, excessive excretion of potassium, high blood pressure and sometimes heart failure.

Cyclamate

Cyclamate is available as sodium or calcium salt of cyclamic acid. The sweetener is easily soluble in water, stable in heat or cold and has extensive shelf life. Cyclamates are 30 times sweeter than sugar, leave little aftertaste and are heat stable.

Cyclamate was banned in the US in 1970 suspected to a carcinogen. Cyclamates have been pending re-approval by the FDA since 1982, and are approved for use in 50 other countries including Canada. Cyclamate was also implicated along with saccharin in causing bladder cancer. It was reviewed and exonerated in 1982, but it is still a banned compound in India. Use of cyclamates in Canada is restricted to table-top sweeteners.

Advantages of using in food products:

- It leaves no bitter taste and has a synergistic sweetness effect with saccharin.
- It is a non-caloric sweetener.

- As they do not mask fruit flavours, it may be of use in food products containing fruit i.e. fruit ice cream, jam, etc.

Disadvantages of using it in food products:

- It can be hydrolysed by gut microflora to cyclohexamine, a known carcinogen.

Stevioside

Stevia rebaudiana is a naturally, sweet herbal plant native to South America, mainly Paraguay and Brazil. Stevia is an all-natural sweetener because it is extracted from the Stevia plant and undergoes no chemical change during the manufacturing process. The leaf extract comprisesmainly of glycosides, namely, stevioside and rebaudioside. These glycosides are 200 and 300 times sweeter then sucrose respectively. Stevia can be used in cooking, baking and frying.

Stevia is available under trade name such as TruVia and PureVia which were patented by Coca Cola and Pepsi. Many different forms of Stevia exist such as Reb A, Reb B, Reb C, Reb D, Rebiana, Stevioside, Sun Crystals and Enliten. Reb A is sold as a tabletop sweetener as PureVia, Truvia, SUN CRYSTALs, and Stevia Extract IN THE RAW. It is marketed in India by the brand name 'Stugar'. Currently, the FDA approves Stevia for general use, with the exclusion of meat and poultry. The FDA approved GRAS status to 95% pure Reb A in December 2008 for its use as a sweetener in food and beverages. The World Health Organization's Joint Expert Committee on Food Additives (JECFA) deemed highly pure Reb A safe for use as a general sweetener in 2008.

The glycosides present in stevioside are not metabolized and pass out unabsorbed from the body. It has not shown any teratogenic, carcinogenic activity in rats. It is approved as a 'dietary supplement' by the FDA. It is used widely in the South America and the Orient. It is popular in Japan, China and Korea and has been used there for more than three decades.

Advantage for use in food products:

- It is completely natural intense sweetener.
- It is heat stable up to 200 °C.
- It is relatively stable under a fairly wide range of pH and can be stored for a relatively long period of time.

Alitame and Neotame

These non-nutritive sweeteners that are awaiting approval from the FDA. They are structurally similar to aspartame. Alitame is a dipeptide based high intensity sweetener formed from the amino acids, L-aspartic and D-alanine, and a novel amine. Alitame is a crystalline, odorless, non-hygroscopic powder. It is partially caloric since the aspartic acid portion of the molecule is available for normal amino acid metabolism. The caloric contribution of 1.4 k.cal/g of alitame is insignicant at use levels in the diet. It only contributes to about 0.02% of the calories of the replaced sucrose, since it is 2000 times sweet that of sucrose. Alitame is synergistic with acesulfame-K and cyclamate and high quality blends may be obtained with these and other sweeteners, including saccharin.

Neotame is chemically similar to the artificial sweetener aspartame, but is used at vastly lower levels and is more stable. Chemically, Neotame has a 3,3-dimethylbutyl group attached to the amino group of the aspartic acid portion of the molecule. Neotame is ~ 8000 times (7,000 to 13,000 times) sweeter than sucrose. It was approved for use as a general purpose sweetener by the FDA in 2002. Neotame is being used in soy-based, nutritionally fortified product. It is mainly used in foods and beverages; other applications include chewing gum, carbonated soft drinks, refrigerated and non-refrigerated ready-to-drink beverages, table top sweeteners, frozen desserts and novelties, puddings and fillings, yogurt-type products, baked goods, and candies. It also can be used in both cooking and baking. Their carcinogenicity and reproductive toxicity reports are yet to be scrutinized.

Commercial Use of Intense Sweeteners

Sweetening substitutes have gained prominence in recent years, catering to a diet conscious market or to consumers who must restrict their sugar intake. Depending on the sucrose replacer chosen in the preparation of a sweetened food product, the products are marketed as 'low-calorie', 'low-cariogenic' or 'dietetic'. Artificial sweeteners are used in products used to limit caloric intake or prevent dental cavities. Sugar alcohols are natural compounds with varying degrees of sweetness which are often added to boost or fine tune flavors of products while increasing their sweetness. They are often used in conjuncture with natural or artificial sweeteners in order to achieve a desired degree of sweetness, taste or texture. Sugar alcohols typically provide some amount of nutrition but have other benefits such as not affecting insulin response or promoting tooth decay, which makes them a popular sweetening choice.

The sensation of sweetness caused by intense sweetener compounds is sometimes notably different from sucrose, so they are often used in complex mixtures that achieve the most natural sweet sensation. In the US, six intensely-sweet sugar substitutes have been approved for use, namely saccharin, aspartame, sucralose, neotame, acesulfame-K and stevia.

Fructose and other non-nutritive sweeteners like saccharin, aspartame, sucralose, acesulfame-K, stevioside are used as table top sweeteners. Sorbitol, sucralose, acesulfame-K are used in baked products as they can withstand high temperature cooking. Confections, candy, lozenges and chewing gum mainly contain polyols like sorbitol, xylitol, lactitol and erythritol. They may also use sucralose as a sweetening agent. Saccharin and aspartame are mainly used in cold beverages, as they cannot withstand high temperatures. Other sweeteners like sucralose, fructose, polyols and stevioside are used in a variety of beverages, both hot and cold.

Application of Sweetener in Diabetic Foods

Diabetic foods mainly contain sorbitol, sucralose and saccharin. Polyols and fructo-fibres essentially are used a bulking agents or diluents with many sweeteners.

Chemistry of Food Flavors

Flavor is caused by receptors in the mouth and nose detecting chemicals found within food. These receptors respond by producing signals that are interpreted by the brain as sensations of taste and aroma. Certain taste and aroma combinations are characteristic of particular foods.

Aroma in Food

The term aroma includes the global integral perception of all the senses that are involved (smell, taste, sight, touch) in consuming food.

There are almost unlimited variations in the intensity and quality of odours that can be perceived by the specialised cells of the olfactory epithelium from the nasal cavity. The gustatory papillae situated on the (top of the) tongue and on the posterior part of the oral cavity facilitate the perception of sensations of sweetness, sharpness, saltiness and bitterness. The detection of sensations of spiciness, coldness and umami is realised by the cells from the trigeminus.

The aromatic substances present a wide range of chemical structures that stem from the main constituents of food. They are characterised by the fact that they stimulate the receptors of smell and/or taste in order to produce an integrated psychological response.

Savory Substances: Sweet, Bitter, Sharp and Salty

There is a theory that relates the molecular structure to the sweet flavour and satisfactorily explains the reason why certain compounds have a sweetening capacity.

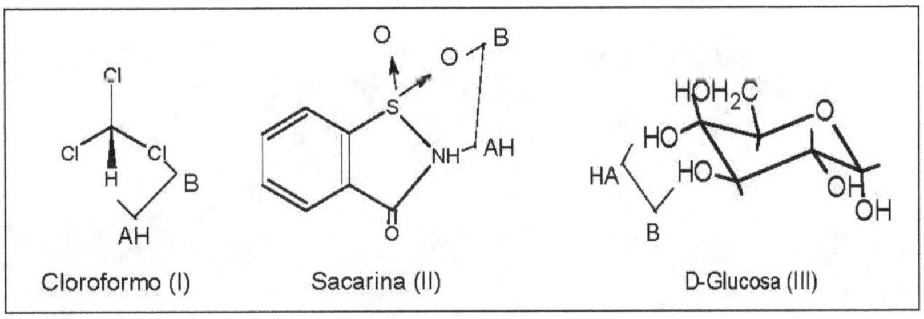

AH/B relations for structures of chloroform (I), saccharin (II), and D-glucose (III).

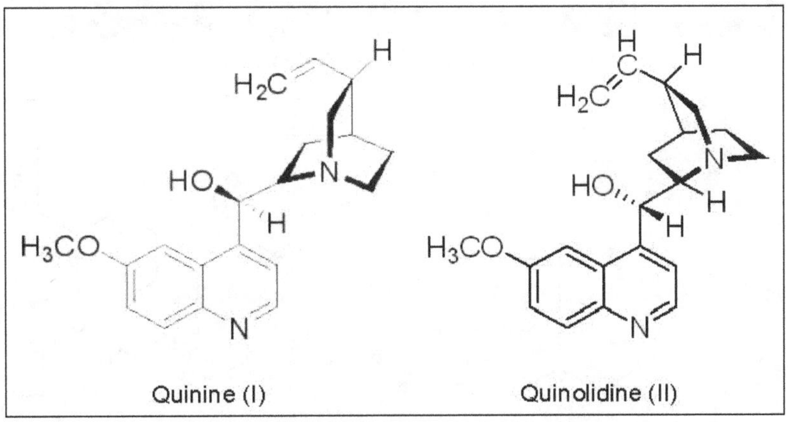

Quinine and quinolidine chemical structure.

Bitterness is similar to sweetness since it depends on the stereochemistry of the molecules that generate the stimulus. The quinine, an stereoisomer of quinidine, is an alkaloid accepted as the standard substance that represents the bitter gustatory sensation. This natural alkaloid, of white and crystalline colour, is extracted from the bark of the cinchona, an American tree belonging to the family Rubiaceae that presents antipyretic, antimalarial and analgesic properties. Due to the side effects of high doses of quinine, its concentration has been limited by the legislations of each country.

Quinine tree

The substance responsible for the bitter flavour of fruit, especially in the cases of the grapefruit, navel orange, Valencia orange is a triterpenoid dilactone denominated limonin. This is the substance which actually gives citrus fruits their characteristic sharpness. The limonin is a natural bactericide whose main function is to prevent the plant from contracting diseases.

Grapefruit, navel orange and Valencia orange.

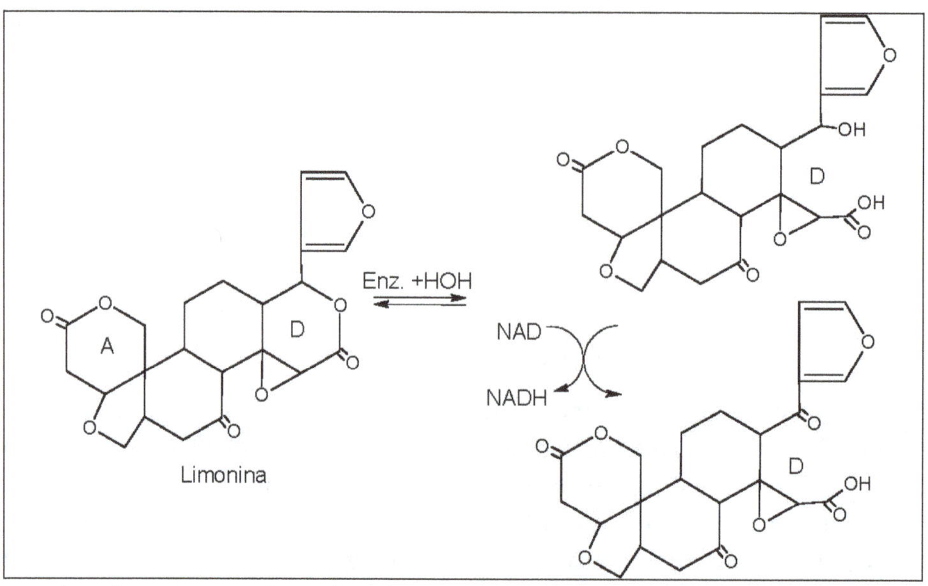

Limonin structure and reactions that enzymatically lead to non-bitter derivatives.

The naringin is a flavonoid (flavanona glicosilada), present in grapefruit and bitter oranges, which are extracted from the rind of some citruses (Citrus paradise, Citrus aurantium, Citrus maxima). The narangin is the main responsible for their bitter flavour. It can also be found in fruit flesh, leaves, flowers and plant seeds. The narangin is used in the perfume industry, or to give flavour to sweets, drinks and bakery products. It is still being studied for its antioxidant property, as an oil stabilizer, an anti-mutagenic agent, and as a precursor of the compound naringin dihydrochalcone due to its great sweetening capacity. Salty and sharp flavours are related to salts, although they have complex flavours that consist of psychological mixtures of sweet, bitter, sharp and salty perception components.

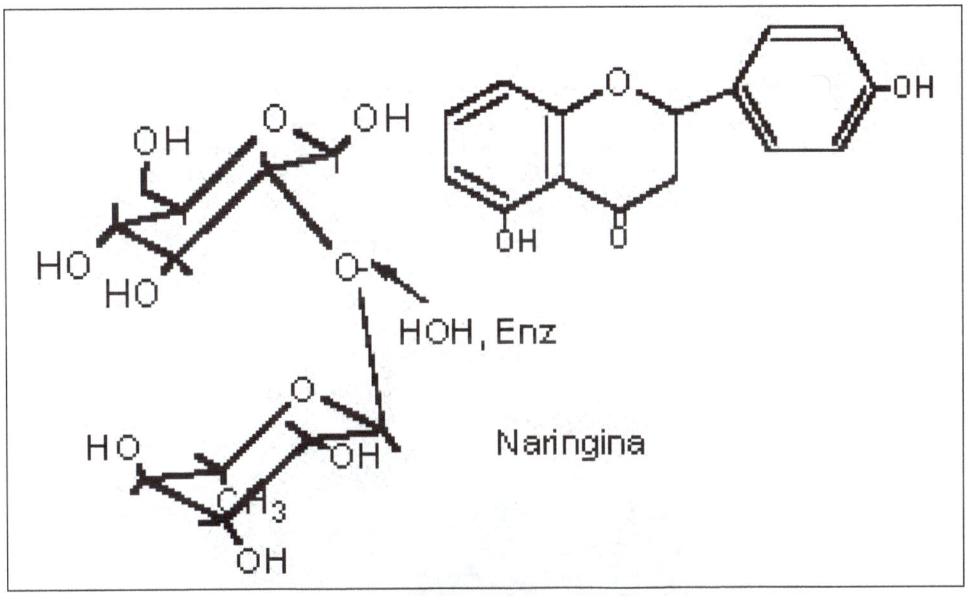

The structure of the narangin showing the step of the enzymatic hydrolysis that leads to non-bitter derivatives.

Citrus paradisi, Citrus aurantium, Citrus maxima.

Aroma Enhancers

These substances contribute to the umami or delicious food flavour when they are used at levels that exceed their own threshold of detection, increasing the aromas to levels below their own thresholds of detection. The most known representatives of this group are the monosodic L- glutamate (MSG) and the 5`-inosine monophosphate (5`IMP).

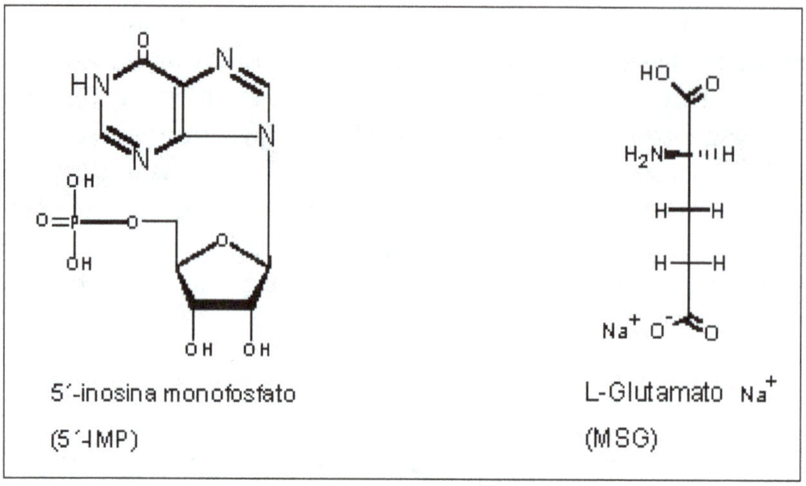

Structures of 5`-inosine monophosphate (IMP) and sodic L- glutamate (MSG).

Laminaria Japonica

The monosodic L-glutamate (MSG), sodium salt of the amino acid known as glutamic acid can be naturally found in various foods like tomatoes, mushrooms, and including maternal milk. The monosodic L-glutamate is a nutritious additive used as a flavour enhancer. First, the MSG was extracted from the alga Laminaria japonica. The MSG is produced through the fermentation of natural products such as the treacle of the sugar cane or cereals. They are fermented in a controlled environment by using microorganisms and then filtered and purified until refined monosodic glutamate is obtained.

The monophosphate inosine (IMP) is an intermediary nucleoside of the pathway of the nucleic acid synthesis, basic cell components. It is a precursor of the adenosine, a molecule which is important for energy production and has a great number of support functions in the organism. The inosine is found in the brewer´s yeast and in animal entrails. The inosine is also available in purified form as well as in supplement. At industrial scale, the inosine is primarily used for its addition to prepared foods, instant soups, etc., since it has a great flavour enhancing effect.

Saccharomyces cerevisiae (brewer´s yeast) and Corynebacterium glutamicum.

Astringency

Astringency is a phenomenon associated with flavour due to the association of tannins and polyphenols with proteins in the saliva forming precipitates and aggregates.

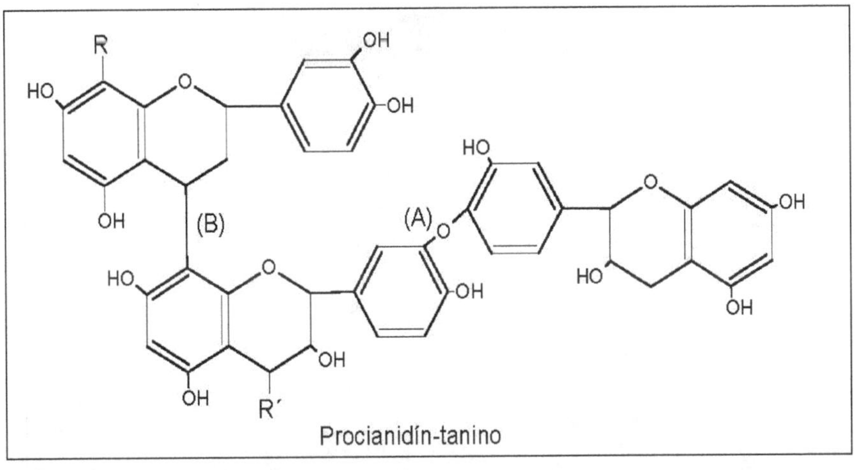

Structure of a procyadinin-tannin showing the condensed tannin bond (B) and the hydrolisible tannin bond (A); there can also be noticed the great hydrophobic areas that can associate with proteins and produce astringency.

Tannins are very astringent polyphenolic compounds of a bitter flavour. Industrially, they have been used for their properties of tanning the skin on having eliminated the water from the muscular fibres. The chestnut tree (Castanea sativa) produces a hydrolysable tannin which is used in the leather industry. The importance of tannins is also given by their capacity of protecting plants from injuries and protecting themselves against outside attacks because they result toxic for microorganisms and herbivores, and they are not digestible for the former group. Their flavour is very sour and they produce dryness in the mucous membrane of the mouth when eating them. This capacity of drying the mucous membranes is what causes the astringency sensation.

Polyphenols are a heterogenous set of molecules that share the characteristic of having in their structure various benzene structures replaced by hydroxylic functions. The phenolic compounds are very important in food products like wine, especially the coumaric acid, cinnamic acid, caffeic acid, gentisic acid and ferulic acid that are synthetized through a common metabolic pathway starting from the phenylalanine. The polyphenols are important in the plants´ physiology because they contribute to the resistence to microorganisms and insects and help to preserve their integrity during their continuous exposure to stressful environments, including ultraviolet radiations and relatively high temperatures. In the human body, part of the biological activity of the polyphenols is very important due to the capacity of being part of the cellular antioxidant system.

Castanea sativa

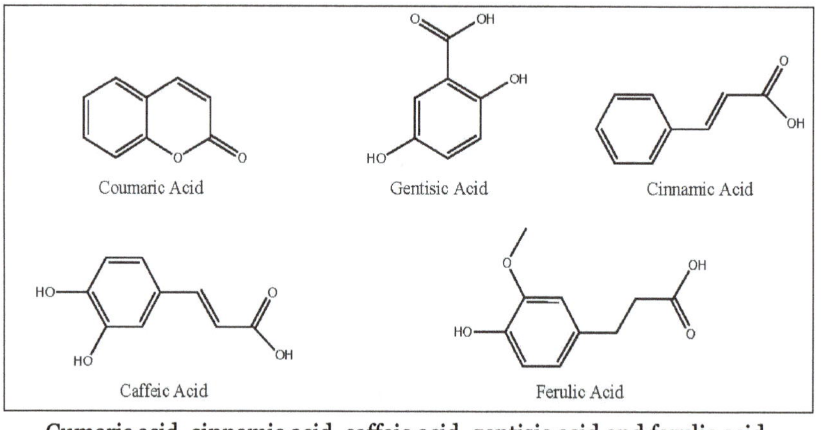

Cumaric acid, cinnamic acid, caffeic acid, gentisic acid and ferulic acid.

Spicy Effect

The main piquant spices of alimentary use are: Chili pepper, ginger and pepper.

The capsaicin (8-methyl-N-vanillyl-6-nonenamide) is the active component of the chili pepper (Capsicum). The capsaicin and other related substances are called capsaicinoids and are produced as a secondary metabolite in various plants of the genus Capsicum, probably in order to avoid being eaten by herbivores. It is an irritant for mammals producing a strong sensation of burning in the mouth. The pure capsaicin is a compound characterised by its lipophilic character and is an odourless, colourless and wax-like substance.

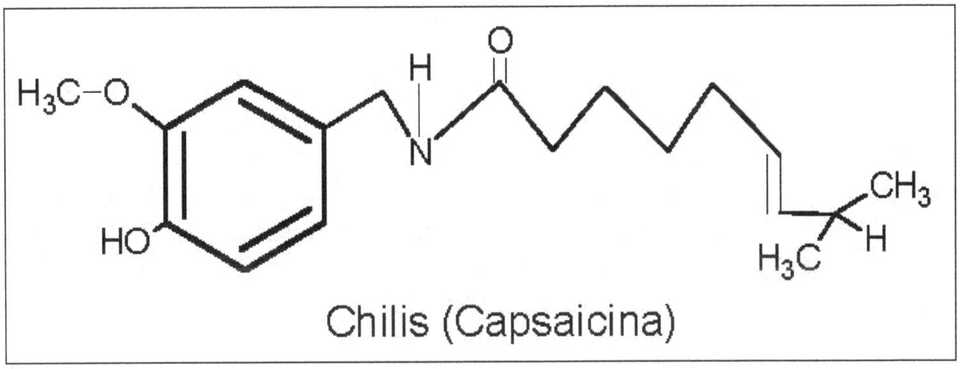

Chili structure

The piperine is the most important piquant component of the pepper and was first isolated in 1820. The piperin is the main alkamide present in the Piper nigrum, and even if it is accompanied by many other alkaloids, there can be observed isobutylamides with isometry different to that of the affinine. The alkamides present in the fruits are the main responsibles for the spicy flavour of pepper, although its flavour is also characterised by the great variety of amide structures and the considerable range of alkaloids. The clarified oil of pepper which is commercially available contains 55% of piperine.

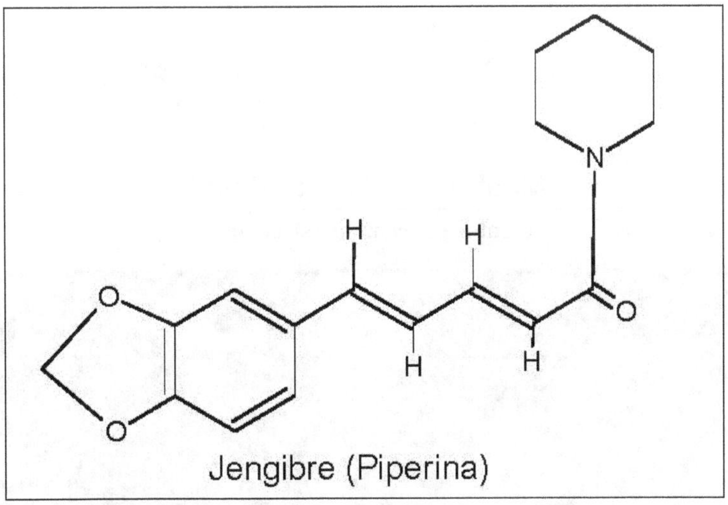

Ginger structure

The spicy flavour of ginger is due to gingerol, an oily liquid constituted by equivalent phenols. The main one is gingerol. It is formed in the plant from phenylalanine, malonate and hexanoate. Gingerols of different length of the lateral hydrocarbonated chain have been established by the Japanese investigators; other minor components of an extract are gingediols, methylgingediol, gingediacetates, and methyl gingediacetates.

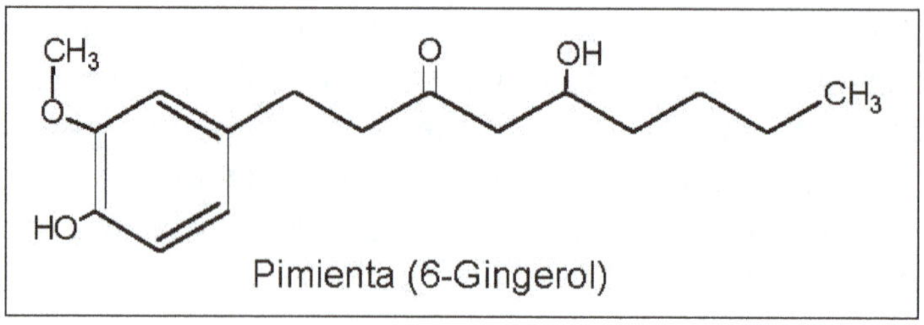

Pepper structure

Refreshing Effect

Two of the most used compounds that produce this sensation are the menthol and the camphor. The menthol of plant origin is the product of the natural distillation of mint leaves. Its composition is identical to that of the mint, but of a higher concentration of elements. The menthol is used for sweets (pills, chewing gum, etc.). It is also used as a medicine for treating respiratory diseases. It is an intestinal antiseptic, anaesthetic and decongestant.

Camphor is a semisold, crystalline and waxy substance with a strong and sharp acrid odour. It is a terpenoid with the chemical formula $C_{10}H_{16}O$ and it is found in the wood of the Cinnamomum camphora, known as the camphor tree, a large evergreen tree orginated in Asia. It is used as a balsam and for other medicinal purposes. In large quantities, it is poisonous when ingested and can cause seizures, confusion, irritability and neuromuscular hyperactivity.

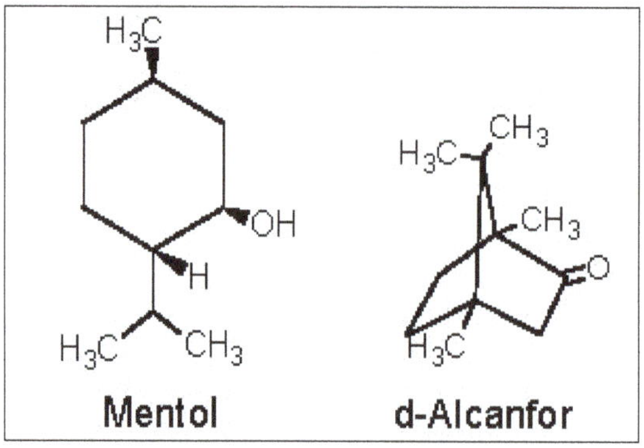

Mentol and camphor structures.

Mint and Cinnamomum camphora plants.

Aroma of Vegetables, Fruits, and Spices

The volatile thioderivatives are the precursors of the characteristic aroma of the genus Allium (onion, garlic, leek, chives, and shallot), and of the genus Brassica of the Cruciferae (cabbage, turnip, Indian/brown/leaf mustard, watercress, radish). The volatile methoxy alkyl pyrazines are responsible for the brownish-green aroma that many vegetables possess.

Fatty acids are high-energy molecules necessary for all the cellular processes in the presence of oxygen because their content of hydrogens allows them to oxidize more than glucides or other non-reduced organic compounds. These molecules play an important part in the characteristic aroma of fruits and vegetables.

Volatile Terpenoids Responsible for Aroma

The terpenes are the major responsibles for the aroma in citrus fruits, spices, and aromatic plants. The monoterpenes contain ten carbon atoms, the sesquiterpenes contain fifteen. The terpenes constitute some of the ramified alkylic compounds of the natural aroma volatiles, but can be turned into annular compounds that are aromatized afterwards. The oxygenated terpenes exhibit a more pleasant aroma than the nonoxygenated terpenes.

Citral and limonene exhibit aromas of lemon and lime, respectively. The citral belongs to the aldehyde group and is a molecule of fresh, lemon-like, vaporous odour and flavour. It is found in lemons, citronella, oranges and lemon balm. The limonene belongs to the terpene group and is a natural substance that is extracted from the citrus fruits. It is the substance that gives the specific odour to oranges and lemons. Industrially, it is used to produce cymene for catalytic dehydrogenation. In the last years, there has been an increasing demand for its use as a biodegradable solvent.

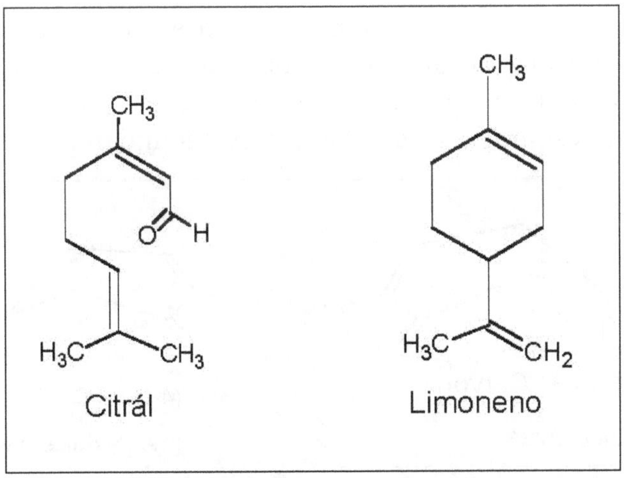

Citral and limonene structure.

The limonene possesses a chiral centre; therefore, there are two optical isomers: D-limonene and L-limonene. The limonene is considered a transfer agent of clean and environmentally innocuous heat so it is used in many processes by the pharmaceutical and alimentary industries as an aromatic component. It is also used to give flavour, in obtaining artificial flavours of mint, for example, and in the manufacture of sweets and chewing gum.

Lemon and lime

β–sinensal and nootkatone contribute to the characteristic aroma of oranges and grapefruits.

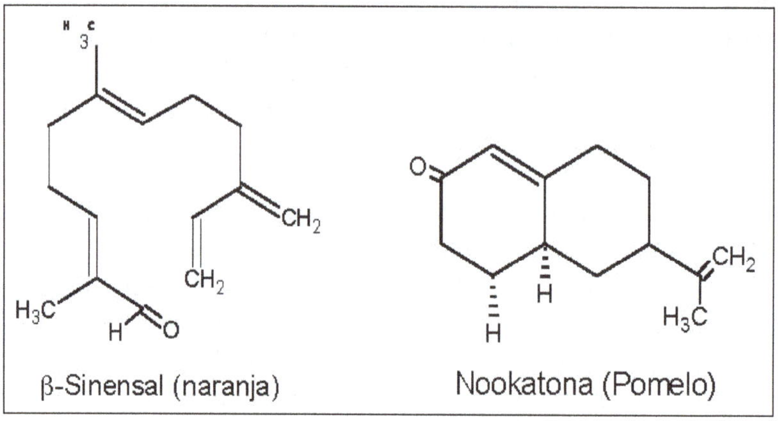

Structure of β-sinensal and nookatone, respectively.

The enantiomeric terpenes may present extreme differences in the fragrance quality. This is the case of L-carvone and D-carvone. D-carvone is a molecule of spicy, herbaceous and sweet odour and flavour. It is found naturally in dill and caraway. As for the L-carvone, it is a molecule of sweet, mentholated, spicy and refreshing odour and flavour. It is found in curly mint.

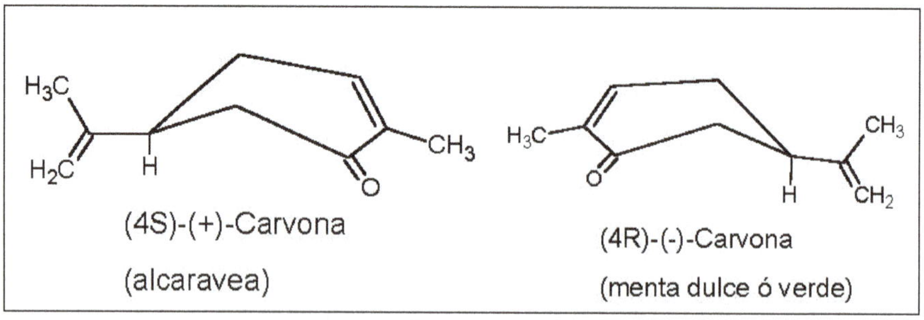

L-carvone and D-carvone structure.

Aromas Derived from the Shikimic Acid Pathway

This pathway, along with the compounds responsible for the aroma derived from the aromatic amino acids, supply volatile compounds related to essential oils and provides the phenylpropanoid structure to the lignin polymers.

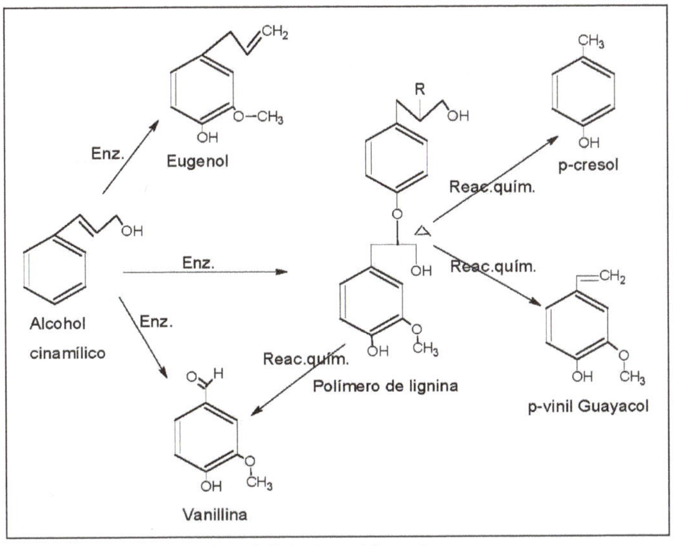

The shikimic acid pathway.

Through this pathway vanillin can be obtained. Vanillin (3-Methoxy4hydroxy benzaldehyde) is one of the most apreciated flavouring substances used to create artificial aromas.

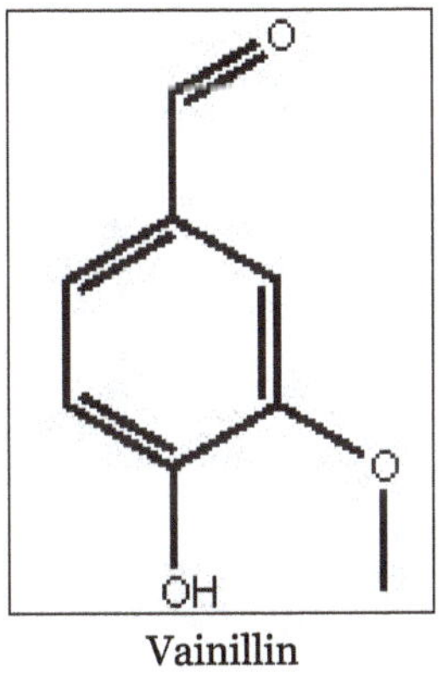

Vainillin

Vanilla is an elaborate flavouring essence which uses the seed pod of the Vanilla orchid. The natural vanilla is an extremely complex mixture of several hundred different compounds, while the synthetic one derives from phenol and high purity. Nevertheless, it is very hard to distinguish any difference between the two.

Vanilla orchid

References

- Chemistry-in-food-flavours-236008821: researchgate.net, Retrieved 28, Jan 2020
- Classification-food-flavours-uses: hmhub.me, Retrieved 17, June 2020
- Food-colors-importance: foodscintech.blogspot.com, Retrieved 11, March 2020
- Sweeteners-for-food, food-additives, food-technology: epgp.inflibnet.ac.in, Retrieved 16, May 2020
- Natural-flavours-obtained-by-microbiological-pathway, generation-of-aromas-and-flavours: intechopen.com, Retrieved 13, Feb 2020

3

Food Emulsifiers and Stabilizers

An additive which facilitates the mixing of two liquids is termed as an emulsifier. Stabilizer is a type additive that helps to preserve the structure of food to which it is added. The topics elaborated in this chapter will help in gaining a better perspective about the common food emulsifiers and food stabilizers.

Emulsifiers are additives that allow normally immiscible liquids, such as oil and water, to form a stable mixture. An emulsion is a blend of two immiscible liquids, with droplets of one phase (dispersed phase) distributed in the other phase (continuous phase). In general these are unstable systems in which the dispersed phase droplets tend to agglomerate or coalescence and separate out. The boundary between the two phases is called the interface.

Emulsifiers contain both hydrophilic and lipophilic parts. The hydrophilic head is easy to hydrate (water soluble) and the lipophilic tail is impossible or very difficult to hydrate (oil soluble). Emulsifiers concentrate at the interface between oil and water and reduce the surface or interfacial tension, thereby making the emulsion more stable. In many industries emulsifiers are referred to as "surfactants", which is an abbreviation of surface active agents.

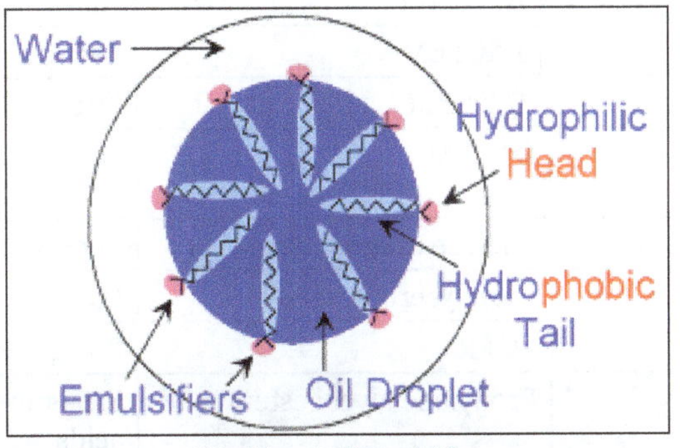

Working of Emulsifier

In commercial food emulsifiers the hydrophilic part can consist of glycerol, sorbitol, sucrose, propylene glycol or polyglycerol. The lipophilic part is formed by fatty acids derived from fats and oils such as soybean oil, rapeseed oil, coconut oil and palm kernel oils. The stability of an emulsion depends on:

- Droplet size: A smaller droplet size facilitates emulsion stability. The droplet size can be influenced by homogenization.
- Viscosity of continuous phase: A higher viscosity facilitates emulsion stability. The viscosity of the water phase can be influenced by the addition of hydrocolloids or thickening agents.
- Specific density of the two phases: If the difference in density between the two phases is small, the emulsion will be more stable. For essential oils in beverages weighting agents can be used to increase the density of the essential oil.

- Quality of the interfacial film: The film can consist of emulsifiers and / or proteins.

HLB: Hydrophilic Lipophilic Balance Concept

Emulsifiers can be characterized by the Hydrophilic Lipophilic Balance. The balance is measured on molecular weight and is an indication of the solubility of the emulsifier. The HLB scale varies between 0 and 20. An emulsifier with a low HLB value is more soluble in oil and promotes water-in-oil emulsions. An emulsifier with a high HLB value is more soluble in water and promotes oil-in-water emulsions.

The HLB value is a somewhat theoretical value, it only considers water and oil, and food systems are more complicated. But the HLB value of an emulsifier can be used as an indication about its possible use.

Emulsifiers Used in the Food Processing Industry

Following emulsifiers are generally used in the food processing industry:

E number	Short name	Long name
E 322	Lecithins	Lecithins
E 432 – 436	Polysorbates	Polyoxyethylene sorbitan esters
E 471	MDG / Monoglycerides	Mono- and diglycerides of fatty acids
E 472 a	ACETEM	Acetic acid esters of MDG
E 472 b	LACTEM	Lactic acid esters of MDG
E 472 c	CITREM	Citric acid esters of MDG
E 472 e	DATEM	Diacetyl tartaric acid esters of MDG
E 473	Sucrose esters	Sucrose esters of fatty acids
E 475	Polyglycerol esters	Polyglycerol esters of fatty acids
E 476	PGPR	Polyglycerol polyricinoleate
E 477	Propylene glycol esters	Propane-1,2-diol esters of fatty acids
E 481	SSL	Sodium stearoyl-2-lactylate
E 482	CSL	Calcium stearoyl-2-lactylate
E 491 – 495		Sorbitan fatty acid esters

Functional Properties of Food Emulsifiers

Functions	Product examples
Emulsification, water-in-oil emulsions	Margarine
Emulsification, oil-in-water emulsions	Mayonaise
Aeration	Whipped toppings
Whippability	Whipped toppings
Inhibition of fat crystallization	Candy

Softening	Candy
Antistaling	Bread
Dough conditioner	Bread dough
Improve loaf volume	Bread
Reduce shortening requirements	Bread
Pan-release agent	Yeast-leavened and other dough and batter products
Fat stabilizer	Food oils
Antispattering agent	Margarine and frying oils
Antisticking agent	Caramel candy
Protective coating	Fresh fruits and vegetables
Surfactant	Molasses
Control viscosity	Molten chocolate
Improved solubility	Instant drinks
Starch complexation	Instant potatoes
Humectant	Cake icings
Plasticizer	Cake icings
Defoaming agent	Sugar production
Stabilization of flavor oils	Flavour emulsions
Promotion of "dryness"	Ice cream
Freeze–thaw stability	Whipped toppings
Improve wetting ability	Instant soups
Inhibition of sugar crystallization	Panned coatings

Emulsifier Selection Criteria

Several parameters should be considered during emulsifier selection. These parameters include:

- Approval of the emulsifier by the appropriate government agency.
- Desired functional properties.
- End product application.
- Processing parameters.
- Synergistic effect of other ingredients.
- Cost.

Desirable Characteristics of Food Emulsifiers

- Ability to reduce surface or interfacial tension.
- Ability to be readily absorbed at the interface.
- Possessing the proper balance at hydrophilic and lipophilic parts (groups), so that desirable type of emulsion (O/W or W/O) can be stabilized.

- Ability to impact a large electro kinetic potential to dispersed phase (droplets).
- Ability to function effectively at low concentration.
- Resistance to chemical changes.
- Lack of odour, colour, etc.
- Safe non-toxic.

Common Food Emulsifiers

Lecithin

Lecithin is widely used in the commercial baking industry. This emulsifier, composed of fatty compounds, is present in eggs. Emulsifying properties are stored in the phospholipids existing in lecithin. It actually prevents the split of water and oil particles.

Good HDL cholesterol content is increased when lecithin is added to the food mixture, as stated by scientists. The droplets of oil in water are kept safe by this emulsifier, increasing the stability and shelf life of the food. Lecithin is overflowing with health benefits too. It prevents high cholesterol and cirrhosis caused by drinking alcohol. Also, it improves nerve, brain and muscle functions.

Fatty Acid Derivatives

There are different emulsifiers that can be derived from fatty acids. To name a few, polyglycerol esters (PGE), polysorbates, stearoyl lactylates, propylene glycol esters (PGMS), and sucrose esters are commonly known. In desserts like cakes and their icings, PGE is famously used. For toppings that are whippable, PGMS is mainly applied. Other products like gums, coffee, sauces, etc need sucrose esters in holding their particles.

Polyglycerol Polyricinoleate (PGPR)

Baking is an appealing activity, especially for moms. We all can't deny that cookies are tasty that's why our sweet tooths would always go for it anytime. In manufacturing chocolates for applications such as baking, polyglycerol polyricinoleate (PGPR) works in enhancing the thickness and volume of the product. Chocolate coatings flow satisfactorily when PGPR is added unto its mixture. It also complements lecithin when combined.

Factories find PGPR as a helpful agent in maintaining the good quality of the chocolate or other products that require certain smoothness and viscosity. With that, baking will be much fun for everybody who loves doing it.

Ammonium Phosphatide (AMP)

Ammonium phosphatide (AMP) is sunflower-based. The use of AMP has been most triumphant in chocolate and confectionery manufacturing. It is chiefly efficacious in achieving uniformity and steadiness of the mixture, leading to high-quality food products. It does pretty well in keeping the right attributes of the food. The size, texture, smell, texture, thickness, etc. AMP can be used as a good alternative of lecithin but it can also be applied with it plus the PGPR.

Mono and Diglycerides

Monoglycerides stay firm in the so-called 'apha crystalline formation'. As it is very versatile, it works well in foams that are whippable while managing the agglomeration of fats. When water molecules need to be dispersed in a fat phase, monoglycerides serve as an instrument that fairly distributes water into the oil.

For products like chocolates, it gives the sensation that feels like the food product is melting inside your mouth – adding the tastiness of the food. It prompts the smoothness and consistency of the processed bulk. The crystalline structure of the food becomes balanced through its help.

Function of Emulsifiers

The use of food emulsifier began with adding monoglyceride and lecithin to margarine. Originally, these substances were only known as emulsifiers. However, as studies progressed, more functions were found and they began to be used in various fields, such as bread, ice cream and cake. Nowadays, emulsifiers are applied to Japanese foods like tofu and minced fish products. Why do bread and tofu need emulsifiers? Generally speaking, an emulsifier is well known for its emulsifying effects, however, actually it has various functions and followings are just some examples:

- Modifies oil crystal and prevents water spattering in cooking.
- Destroys emulsion to stabilize foam and to make smooth texture in ice cream, and keeps its shape.
- Reacts with proteins to make smooth easy-rising dough in bread.
- Acts on starch to make bread soft.

<table>
<tr><th colspan="3">Function of Emulsifiers</th><th>Application Examples</th></tr>
<tr><td rowspan="9">Surface Active Ability</td><td rowspan="2">Emulsification</td><td>W/O</td><td>Margarine, Butter, Butter Cream</td></tr>
<tr><td>O/W</td><td>Ice cream, Cream, Milk Drink</td></tr>
<tr><td colspan="2">Dispersion</td><td>Chocolate, Cocoa, Peanut Butter</td></tr>
<tr><td colspan="2">Foaming</td><td>Cake, Desserts</td></tr>
<tr><td colspan="2">Anti-foaming/Defoaming</td><td>Tofu, Fermentation Industry, Jam</td></tr>
<tr><td colspan="2">Wetting</td><td>Powdered Foods, Chewing Gum</td></tr>
<tr><td colspan="2">Solubilization</td><td>Flavors</td></tr>
<tr><td colspan="2">Cleaning</td><td>Cleaning Agent for Food Industry</td></tr>
<tr style="display:none"></tr>
<tr><td rowspan="4">Starch Complex Forming Ability</td><td colspan="2">Protection of Starch Granule</td><td>Instant Mashed Potato</td></tr>
<tr><td colspan="2">Anti-staling</td><td>Bread, Cakes</td></tr>
<tr><td colspan="2">Prevention of Sticking</td><td>Pasta, Noodles, Rice</td></tr>
<tr><td colspan="2">Prevention of Gelatinization</td><td>Flour Paste, Desserts</td></tr>
</table>

Modifying Ability for Oils and Fats	Crystal Modification	Margarine, Shortening, Chocolate
	Creaming Ability	Margarine, Shortening
	Water-Holding Ability	Margarine, Shortening
	Others	Coating Agent, Lubricant Agent
Protein Modifying Ability	Gluten Modification	Dough Conditioner
	Others	Tofu, Frozen Surimi
Others	Antibacterial and Anti-Fungal	
	Plasticizing	
	Anti-Oxidation	

Emulsification

Oil and water produce emulsion by stirring; however, the emulsion starts to break down immediately after stirring is stopped. The purpose of emulsification is to stabilize emulsion state by preventing break down which occurs due to creaming aggregation and coalescence. To solve these issues, decreasing size of dispersed particles, reducing the density different of dispersion and protecting the surface of oil droplets are effective.

There are two types of emulsion, O/W emulsion or oil droplets in water, which can be found in ice cream and or milk, and W/O emulsion or water droplets in oil, found in butter and margarine. Recently, developments of W/O/W type emulsion or water dispersed within oil droplets of O/W type emulsion and O/W/O type, an opposite type emulsion have been progressing.

These multiple type emulsions not only make low calorie items such as cream which contain less oil, but also stabilize the emulsion by dissolving the unstable substance present in the deepest region of water droplets. Taste can also be enhanced by injecting seasonings and flavors into the water droplets.

Dispersion

Water-insoluble fine powder like cocoa is difficult to disperse because small lumps tend to form on the surface of interface. Powders gradually aggregate and precipitate even after dispersing by shaking. Maintaining suspension of water-insoluble fine powder is called dispersion. When the dispersing material is liquid, we call it emulsion.

Cocoa is produced by dispersing cocoa powders in water (O/W type emulsion), while chocolate is made by dispersing them in oil (W/O type emulsion). Emulsifier adsorbed on the surface of insoluble fine powder changes the particle surface to be hydrophilic or lipophilic. This results in producing water or fats and oils at the outer layer and stabilizing of suspension by an increase in the affinity to water or oil in the outer phase.

Foaming

Foaming ability is one of the major characteristics of emulsifiers. When a solution containing an emulsifier is stirred, the emulsifier is adsorbed on the surface of the produced foam to make a mono-molecular layer and the foam outside of the solution makes a bimolecular layer of the emulsifier. The film coating a bubble is about 100 times thicker than a bimolecular layer, but a bubble breaks off as soon as migration of liquid trapped between bimolecular films occurs.

The addition of emulsifier enables foaming and stabilizes emulsion state of products, thus, smooth texture and expanded volume can be obtained. Typically, emulsifier for ability above is used for cakes, ice cream, moose, whipped topping, etc.

Anti-Foaming/Defoaming

Emulsifier also has anti-foaming and defoaming ability. Anti-foaming or defoaming agents are used in food production where undesirable foaming may occur in the presence of protein, starch etc. Anti-Foaming/Defoaming Agents Characteristics:

- Water insoluble.
- Floatable on the surface because of its small specific gravity.
- Small surface tension and easy spreading on liquid surface.

These characteristics lower the surface tension, and the foam will get thinner. As these agents spread on the liquid surface, all foam will be diminished.

Wetting

Wetting effects of emulsifier moisten the solid surfaces. If a solid material is mixed with an emulsifier or its surface spreads with an emulsifier, the surface then becomes hydrophilic. For example, chewing gum is apt to stick to teeth.

Adhesion to natural teeth does not occur easily because the enamel is hydrophilic and always wet, whereas the adhesion to artificial teeth, occurs easily. We can prevent adhesion by wetting the surface of chewing gum by adding emulsifier.

Bacteriostatic Effects

Monoglycerides of middle chain fatty acid and some polyglycerol esters have specific bacteriostatic effects, and are used as bacteriostatic agents. They are typically used in hot vendor drinks, flour paste, and canned soup.

Action on Starch

Fresh bread is soft, but it becomes elastic and hardens over time. Starch particles consist of spherical shaped amylopectine and amylose. Raw starch called beta-starch is insoluble in water.

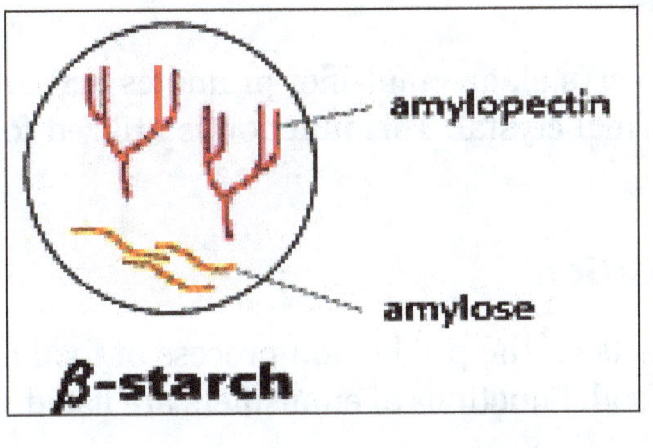

But when heated to a certain temperature, the starch absorbs water and changes into the crystalline form of alpha-starch. Amylose gets hard easily with cooling and amylopectin becomes hard gradually with time. The immediate hardening of bread by cooling is mainly due to the change of amylose.

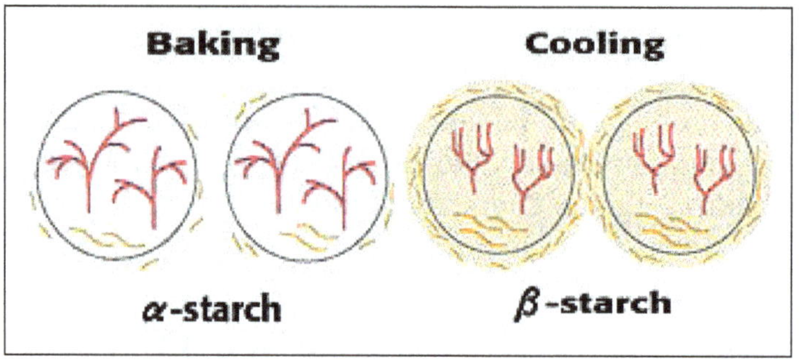

Amylose changes to a helix structure, a kind of spiral structure by heating. If monoglycerides are introduced, they can be anchored into the helix structure. Since the anchoring does not change even after cooling, softness is maintained. This function is also utilized for instant mashed potatoes, noodles and rice.

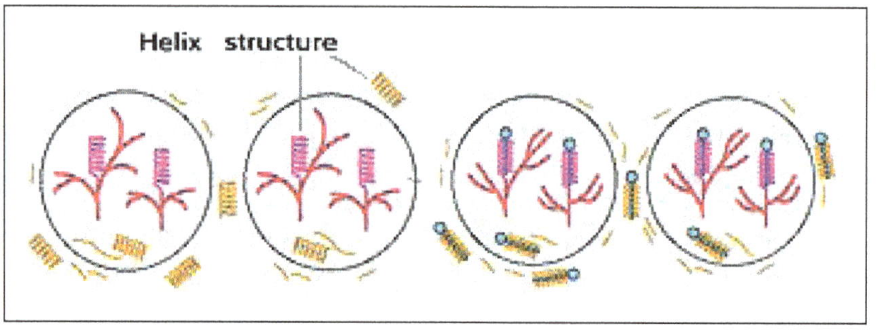

Action on Protein

Wheat flour contains protein called gluten which takes the shape of a mesh-like structured when mixed with water. This wheat flour and water based dough is filled with carbon dioxide produced by fermentation and steam generated during baking, which produces raised bread.

When the gluten content is small, the dough only rises a little. An emulsifier modifies gluten molecules and enhances its film-forming power resulting in good spread ability and improvement of working efficiency. Thus, easy-rising bread can be obtained.

Action on Oils and Fats

When fats and oils form into crystal, an emulsifier promotes and inhibits the growth which prevents the formation of abnormal crystal. This function is utilized for margarine, shortening and chocolate.

Food Emulsifier Application

Emulsifiers have various effects on the production process of food and improve its quality. They are used in various types of food. Functions of emulsifiers are listed.

Bread

Bread and sweet rolls sold at super-markets and convenience stores are usually mass-produced. Mass production and mass distribution require time and speed from start to finish. Emulsifiers are used to maintain the softness as long as possible and to make bread dough suitable for machine production.

Emulsifiers are not only used for emulsification, but also for dough modification, that is, dough gets tolerable against mechanical force by modulating the proteins in wheat flour. Bread is made by mixing and kneading flour, water, salt and yeast, and baking the dough following fermentation.

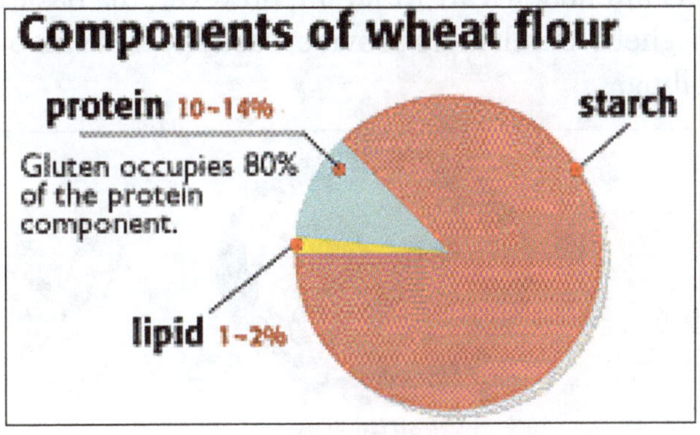

Wheat flour contains 10-14% protein and 1-2% lipid along with starch. 80% of protein is gluten, from which gluten membrane is produced during mixing. Air bubbles are created during mixing, and the gas produced by fermentation expands during baking to fill the space under the gluten membrane. The force due to these gases causes a rising of dough.

At a temperature over 60 °C, starch rapidly expands following water absorption, leading to gelatinization. With a further increase in temperature, the inner phase is made up with denaturation of protein gluten membrane. If this phase is insufficient, gas leaks out and bread expanding does not occur.

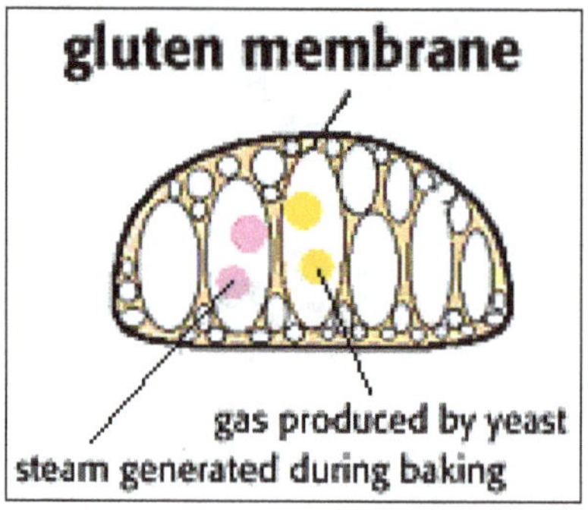

A small amount of lipid contained in wheat flour is said to mediate the formation of gluten membrane. The emulsifier is thought to have similar effects. Major effects of emulsifiers on bread production are as follows:

- To make less-sticky and elastic dough that is easy to handle.
- To increase the dough resistance to mechanical force to soften and create easy raising.
- To make smooth bread of good eating quality, and to maintain the softness.

Noodles and Instant Mashed Potatoes

Emulsifiers make a rigid complex with starch to protect starch granules and improve the quality of starchy foods.

Noodles

Effects of emulsifiers on macaroni, spaghetti of low water content, fresh noodles containing high water content and dried instant noodles are different, however, the basic effects on starch are similar. For macaroni and spaghetti, emulsifiers provide elasticity and smooth uniform surface which prevents sticking after boiling.

In fresh noodles, emulsifiers make easy-to-handle dough and increase the water absorption rate by 1-2%. The surface of noodles becomes smooth, uniform, and less sticky, which improve and streamline the production process. In instant noodles, emulsifiers improve absorption and decrease cooking time.

Instant Mashed Potatoes

The process to make mashed potato are boiling, mashing and drying into flakes. If a starch granule is broken or damaged in this process, amylose in the starch granule is apt to dissolve, resulting in a sticky and glutinous product. When distilled monoglyceride is added just before drying, it forms a starch complex, which protects starch granules. Thus, the mashed potatoes exhibit a smooth texture.

Cakes

Cakes are classified on the basis of the compounding ratio of fat to flour as shown in the following table:

Oils and Fats Content Ranging	
0-10%	Sponge Cake
10-50%	Butter Sponge Cake
30-100%	Butter Cake

Sponge cake is made by utilizing the foaming ability of eggs. Butter sponge cake is made by further additions of butter and margarine. On the other hand butter cake is produced from a batter foamed after the addition of large amount of butter or margarine. Sponge and butter sponge cake are basically made from sugar batter method. Sugar is added to whole eggs after foaming and combined with flour. The use of emulsifier makes it possible to produce an all in one-mixture method, in which all materials are mixed at the same time and foamed.

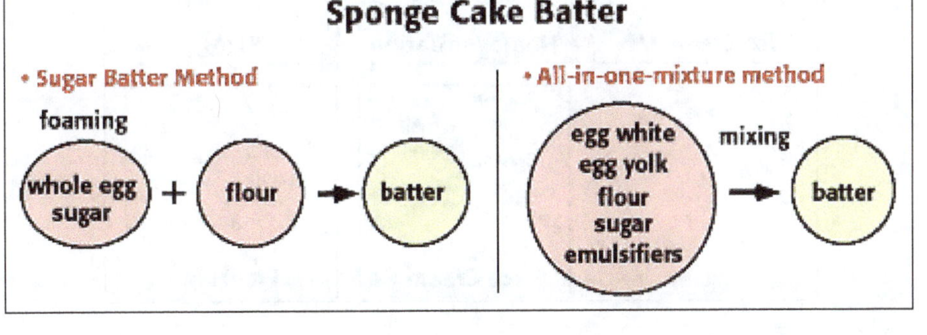

As for the production method for butter cake, there are two methods; sugar batter method, in which foaming is carried out after the addition of sugar to the lipids, and flour batter method, in which flour is added to the lipids before foaming however all-in-one mixture-method becomes possible only by the use of a foaming agent. The roles of emulsifiers in production of cakes are as follows. For sponge cake, fine stable foam is obtainable with it. For butter sponge cake of high fat content or butter cake, emulsifiers control the anti-foaming effects of lipids and keep stable foaming.

In addition:

- Uniform quality products can be continuously produced.
- Big volume, well-balanced products with smooth-surface are obtainable.
- Soft products with homogeneous inner phase are obtainable.

Formulation of Cake

Kinds of Cakes	Wheat Flour	Sugar	Whole Egg	Oils and Fats	Milk
Dorayaki (Bean-Jam Pancakes)	100	70	50	~	~
Sponge Cake	100	100	100	~	~
Sponge Cake	100	120	150	~	~
Nagasaki Castella	100	240	200	~	~
Butter Sponge Cake	100	120	150	25	~
Pound Cake	100	100	100	50	~
Madoreine	100	100	100	100	~
Baumkuchen	100	200	300	100	~
White Butter Cake	100	100	80	40	30
Yellow Cake	100	85	50	45	50
High-Ratio Cake	100	140	59	55	104

Ice Cream

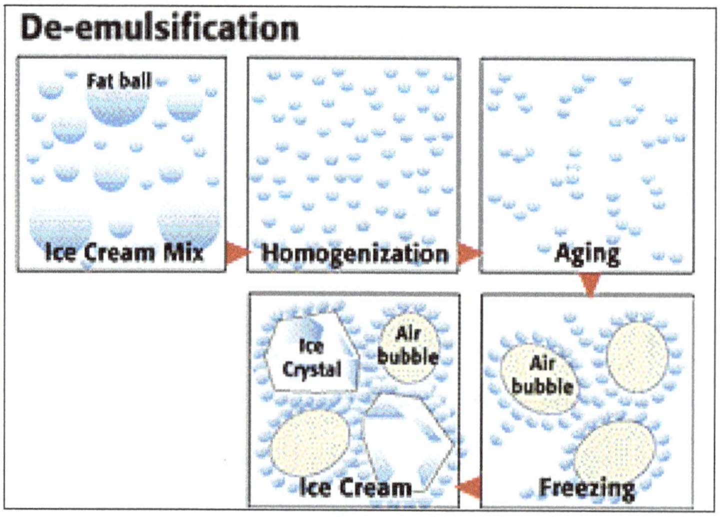

Concerning the manufacturing process of ice cream, a liquid form mixture of raw materials is prepared and homogenized by a high pressure homogenizer. After short period of high-temperature sterilization, the homogenate is stabilized by letting it stand sit for several hours at a temperature below 5 °C. Soft cream is produced by foaming in a freezer at -2 to -9 °C, and pouring into a small container followed by rapid freezing.

Oils and fats, ones of the major components of ice cream, form fat balls by emulsification, and their surfaces are covered with emulsifiers and are coated with milk-like protein like casein. By combining with an emulsifier, the fat balls become fine and stable.

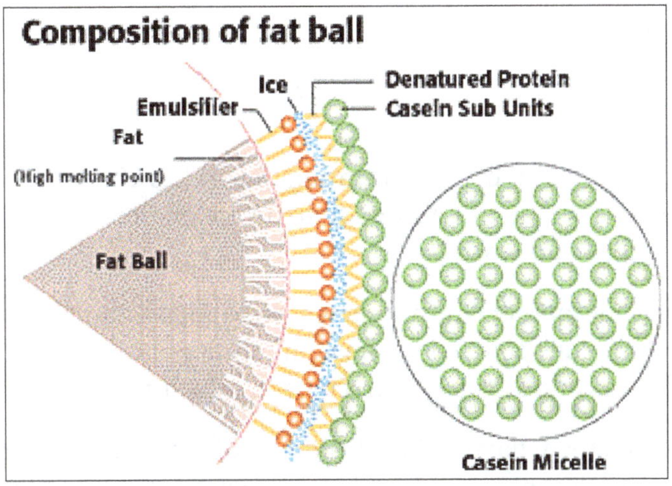

Homogeneous fat balls present in ice cream start to associate in a row during aging, and aggregate like a bunch of grapes. Then, fat balls densely adsorbed on air bubbles surface to make fat ball membranes. Then, a part of fat balls become unstable and the ball surface is broken, resulting in oil droplet aggregations, growth of ice crystals and destruction of the foam. These conditions are called de-emulsification. It is said that these components produced by moderate de-emulsification can be mixed with the foams which is surrounded by stable fat balls to make a skeleton like structure.

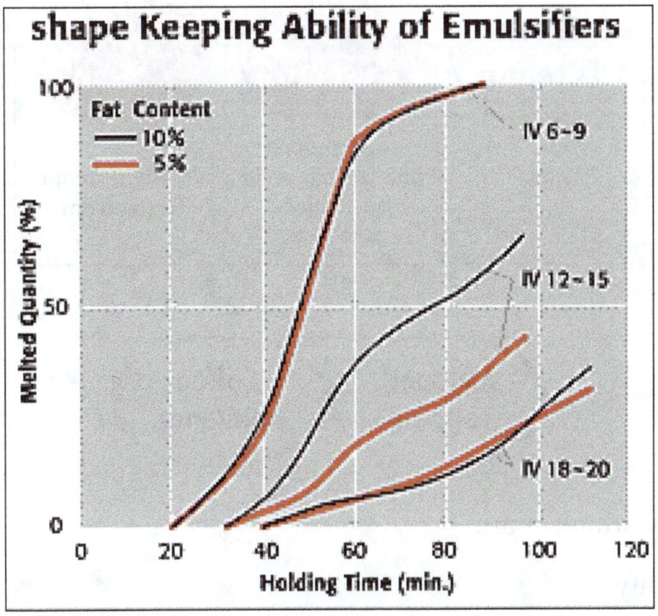

As the degree of de-emulsification varies with the amount of fat used and the kind of emulsifier, it is necessary to select an appropriate combination. The effects of addition of an emulsifier are:

- The progress of uniform fat emulsification is stimulated.
- Stable overrun is obtainable.
- An emulsifier participates in the formation of a dry and firm structure of ice cream leading to smooth texture.
- Heat shock resistance is obtainable.
- Shape keeping ability is enhanced.

Cream

Fresh cream is ordinarily prepared by concentrating milk fat to 18% or more through centrifugation of raw milk. Skilled techniques are needed to foam cream because the quality of fresh cream greatly varies with given feed and seasons.

The foamed fresh cream is unstable and difficult to decorate and easily defoams. Imitation cream was developed by adding non-fat milk solid component, emulsifier and stabilizer to vegetable oil, and it rapidly spread nationwide because of its two characteristics which are pure white color and light flavor.

Furthermore, a compound cream having the advantages of both fresh cream and imitation cream, has been put on the market. Recently, good quality compound cream has become a major product. Cream is a typical product of O/W emulsion.

Representative cream products are coffee whitener and whipped topping. The fat content of coffee whitener is low because the emulsion is more easily broken as coffee is an acidic solution and usually served at high temperature.

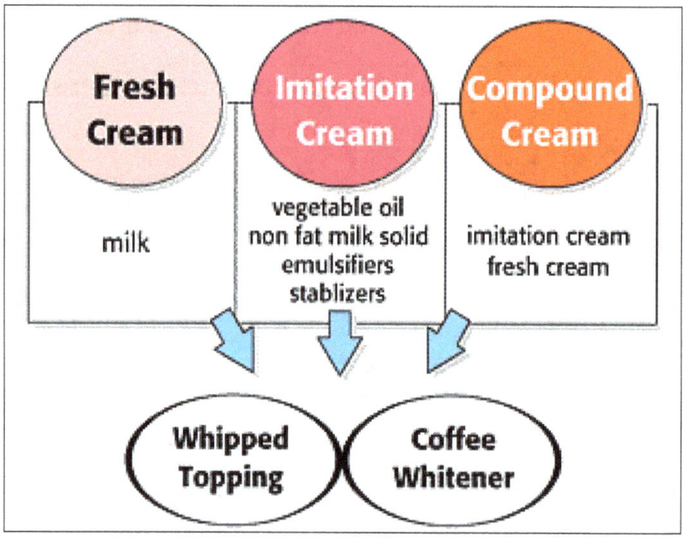

Requirements for coffee whitener are:

- High dispersibility

- Nonoil-separable
- Non feathering by aggregation of milk protein
- Whitening effects

Both emulsification and de-emulsification effects are needed for an emulsifier of whipped Topping. Emulsifier helps the formation of coating membrane from the protein complex on the emulsification of whipped topping so as to make a strong coating membrane sufficiently resistant to high temperature sterilization and cooling.

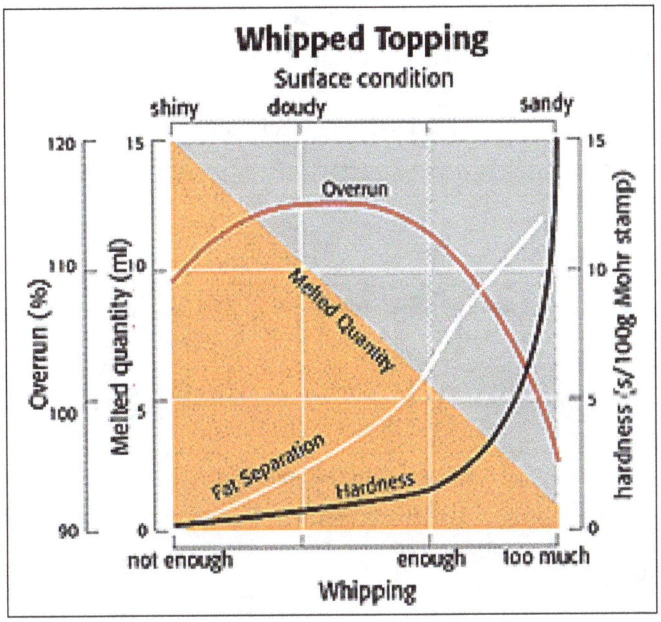

The strongly-stabilized fat balls adsorb on the surface of air bubbles to make a network, resulting in the stabilization of foam. And de-emulsification creates the skeleton like structure of cream by coagulating oil drops in the emulsion and affects taste and shape retention. Accordingly, it is necessary to investigate the properties and the amount of the emulsifier to be used.

Tofu

It is said that tofu was introduced from China and the basic manufacturing method has not been changed since.

Foam Extinguisher

- Anti-foaming effects,
- Defoaming effects.

Increased Protein Content in Soybean Milk

To make cotton tofu, GO has to be created first by grinding with ten times more amount of water than that of soybeans right after soaking soybeans in water. GO is to be separated into soymilk and bean-curd(okara) lees after boiling, modifying protein and filtration. After cooling, a coagulating agent is added to soybean milk and the mixture is poured into a box covered with a cloth. The aqueous part called "Lyu", is removed by weight-loading. The remains are molded. Silken tofu is produced by making concentrated soybean milk with about 5 times the amount of water and standing alone for coagulation.

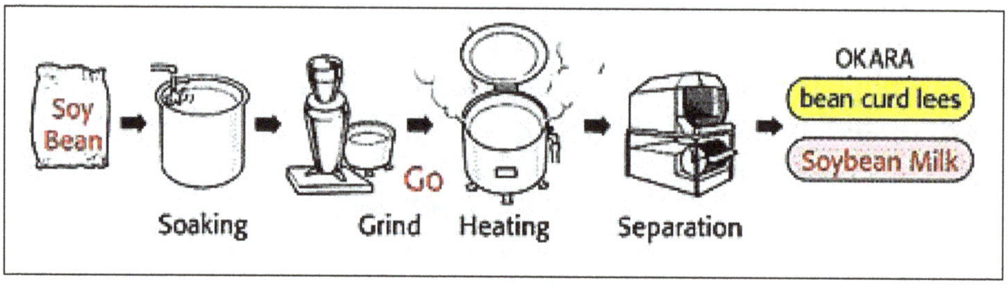

Production flow for Tofu

Cotton Tofu (Momen Tofu)

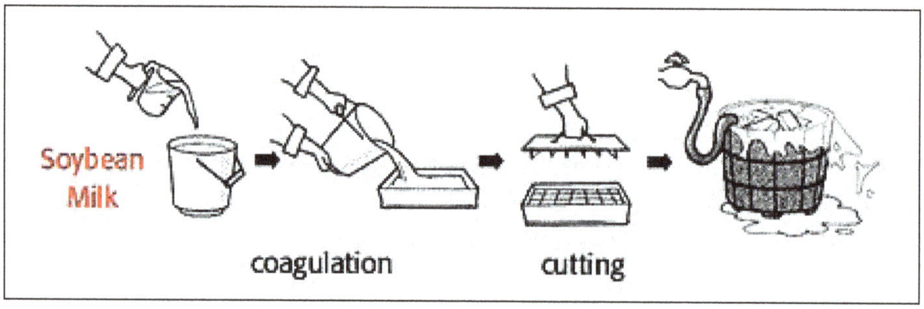

Silken Tofu (Kinugoshi Tofu)

When GO is steamed, its vigorous bubbling occurs in the presence of protein and saponin of soybean by boiling. In a continuous steaming system, the bubbling lowers the heat conductivity, resulting in non-uniformity. Also, numerous bubbles are produced during the separation of bean-curd lees, and the bubbles left in the soybean milk weaken the structure of tofu. Tofu has been produced in the cottage industry and foam extinguisher oil was used as an anti-foam agent. Recently, however, monoglyceride is widely used because of a large scaled manufacturing and spread of hygienic concerns among the public.

Foam extinguisher for tofu has anti-foaming effects to repress foaming as well as defoaming effects. Generally, a distilled monoglyceride of saturated fatty acid is added to "GO" to repress foaming. Other than the anti-foaming effect, monoglyceride stimulates emulsification of "GO", and makes a complex with starch which inhibits swelling, resulting in improved separation of bean-curd lees and increased protein content in the soybean milk.

Margarine and Shortening

Margarine, produced by emulsifying milk with lipids, was born in 1869 in France as a substitute for butter. Some problems with margarine are; separating water droplets and spattering during cooking. Since the 1920s, monoglycerides and lecithin have been used so exclude these defaults.

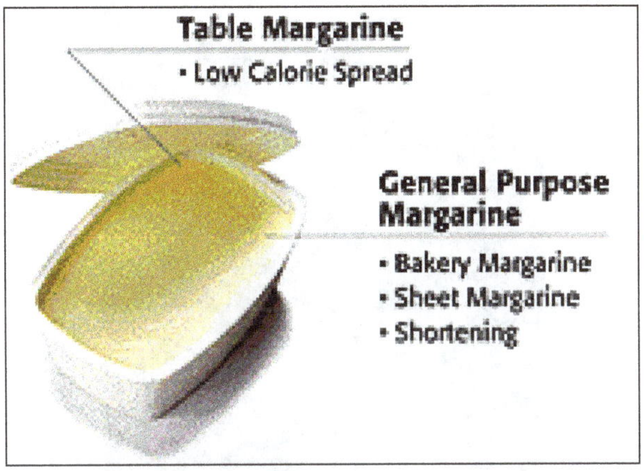

Margarine and shortening are the representatives of processed oils fats. Margarine and low calorie spreads are used in food such as bread, cake, pie, Danish pastry and butter cream, for both household and industrial uses. In addition, there are various functional oils and fats such as fats for fresh and ice creams, powdered fats and releasing oil. For the development of a new processed fat product, it is important to know how to use emulsifiers effectively.

The effects of emulsifiers on processed oils and fats are:

- Emulsification in production process and stabilization of the product.
- Enhancement of spreading effects of oils and fats on bread or pie base.
- Improvement of foaming and air-holding abilities and creaminess.
- Improvement of water retaining ability.
- Improvement of the functions of starch and protein.
- Prevention of spattering during cooking, etc.

In addition to these, emulsifiers have many others function due to the effects on crystal forms of oils and fats as follows.

Prevention of Crystallization

The process to fractionate a liquid fat from solid oil by temperature control is called wintering. Winterized salad oil is a clear liquid, which occasionally changes to a turbid one with time, owing to crystallization. This change is prevented by adding a specific emulsifier. An emulsifier of which solubility in fats and oils is low, deposits as crystals together with fats of low melting point, but these crystals are very fine so that the solution appears to be clear.

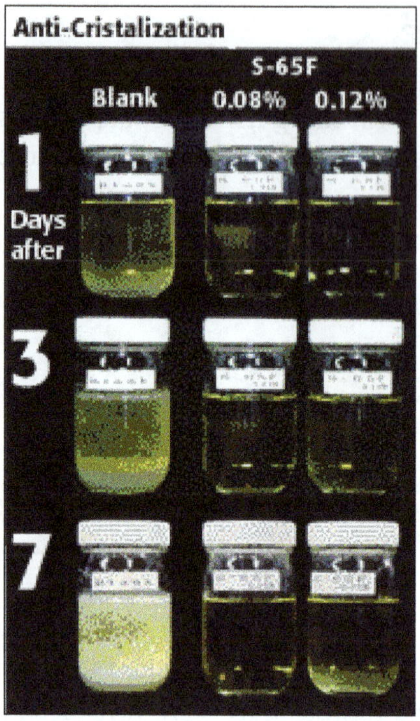

Inhibition of Crystal Growth

When a once-melted chocolate solidifies again, a rough surface of fine powders may be produced. This condition is called blooming. It is attributed to the high crystalline nature of cacao fats, of which triglycerides have comparatively uniform structure. Blooming is also prevented by adding a specific emulsifier. This is caused by the inhibition of crystal growth, which occurs when an emulsifier with a quite different structure is introduced to cacao fats, which have a simple structure.

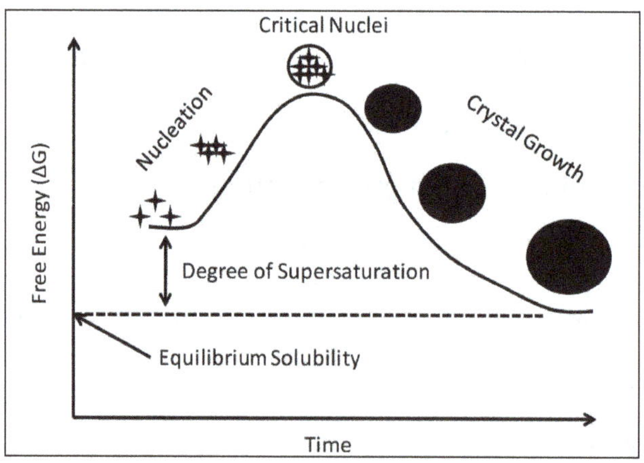

Acceleration of Crystal Growth

Solid fat produced while wintering is removed by filtration. The larger crystals in solid fat are, of course, easily removed by filtration. It is said that when a specific emulsifier is present in the wintering process, the emulsifier functions as a core of the crystal, resulting in the formation of large crystal.

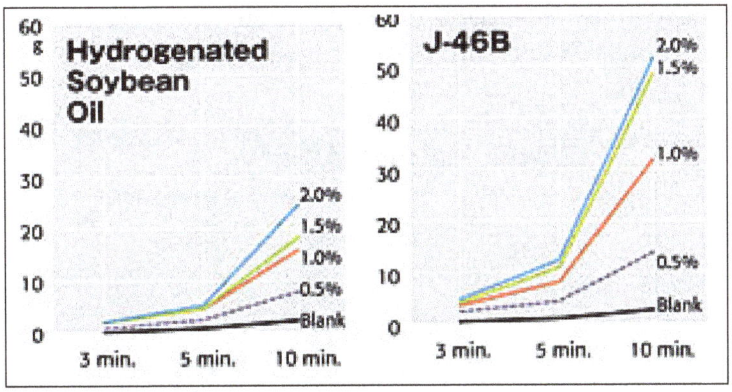

Improve the Plasticity of Fats and Oils

The requirement of processed oils and fats is plasticity regardless of temperatures as seen in margarine which spread easily, even at low temperatures and dose not melt at high temperature. Margarine of which hardness is less variable with temperature was once produced using acetylated monoglycerides or saturated monoglyceride, the crystalline conformation of which holds as a liquid oil.

Soft margarine contains much liquid oil, however, it retains its structure by forming abundant crystals of solid fats. An emulsifier improves the shape keeping capacity of soft margarine by the formation of these abundant crystals of solid fat.

Improvement of Creaming Texture of Fats and Oils

The creamy property enabled through aeration during the mixing process is important for confectionery. Fats and oils used in such items as butter cream and biscuits are affected by crystal forms of fats and oils. Monoglyceride, especially unsaturated monoglyceride like safflower and sunflower monoglyceride, has strong effects on improving creaminess.

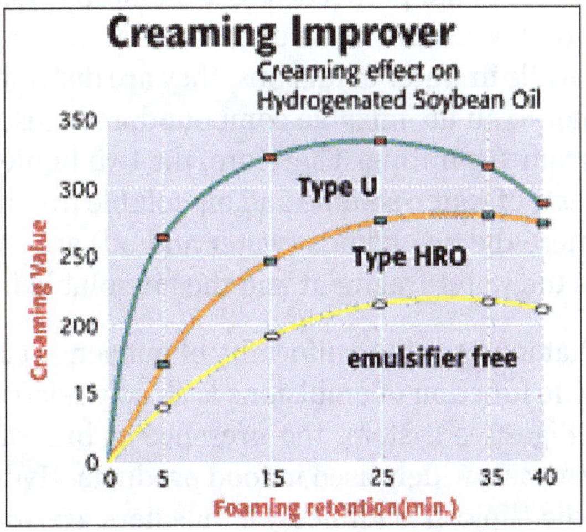

Increase Water-Holding Ability of Fats and Oils

Plastic fats and oils have a water-holding ability, but emulsifiers further enhance it. When increasing amount of emulsifier, holding water ability rises. When an emulsifier is added at 5%, 1 to 2-fold amounts of water can be held. Especially for monoglyceride, when the unsaturation degree is high, the holding-water rate is high. When safflowers and sunflower monoglyceride is added at 10%, water can be held over 5-fold.

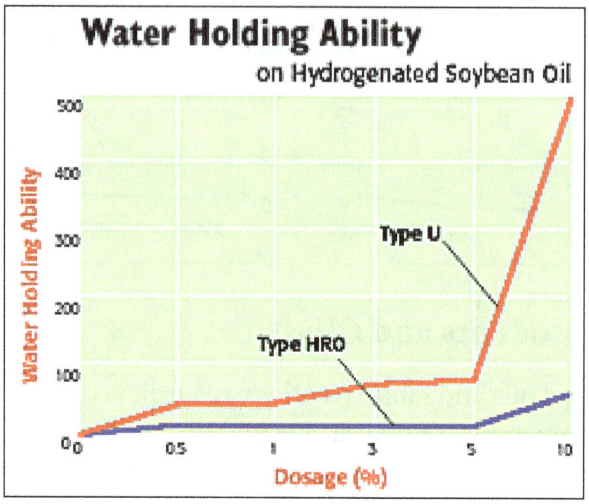

In this case, W/O emulsion is formed. Unsaturated monoglycerides specially produce strong, liquid crystal gel at room temperature, which appears to protect the water drops. Fats and oils for butter cream and icing to be combined with large amounts of syrup are an example of such monoglycerides.

Bioemulsifiers

Bioemulsifiers are higher in molecular weight compared to biosurfactants since they are complex mixtures of heteropolysaccharides, lipopolysaccharides, lipoproteins and proteins. Emulsifiers have double lipophilic and hydrophilic properties. On the other hand, emulsions are either oil-in-water (O/W) or water-in-oil (W/O). In oil emulsions, small droplets of oil form the dispersed phase and discrete in water, while in water emulsions, they are distributed as small droplets of water in oil. Adding an emulsifier to an un-mixable compound, reduces surface tension between the two phases and prevents it from separating. Therefore, the two liquids are able to form an emulsion. Since an emulsion consists of water-soluble and oil-soluble fragments, an emulsifier is placed on the surface of the area where the two liquids (water and oil) are connected. The water-soluble fragment ambulates towards the water fragment and the fat-soluble fragment places near the oil.

Emulsifiers are substances that increase the uniformity of nutrients, such as fatty acids, fat-soluble vitamins, and amino acids. The function of emulsions is closely related to its chemical structure.9 Physiologically, in animal's digestive system, the presence of bile salts, benefits fat absorption. Emulsifiers are surfactant materials widely used in food products. Hydrophilic characteristics (water-friendliness) and lipophilic (lipid-friendliness) emulsifiers are sometimes referred as hydrophilic/lipophilic equilibrium (HLB), indicating the rate of emulsifier's inclination towards water or oil.14-16 Emulsifiers are embedded in fat droplets and prevent the protein layer from collapsing. Other functions of emulsifiers in the food industry include:

- Starch reaction: Most emulsifiers have a lean fatty acid layer in their molecule which form an amylose mixture. This feature is very important in delaying bread and bakery products staling and reducing their adhesion to staple products such as potato puree and pasta.

- Generating interactions with proteins: Emulsifiers have ionic structure which react with proteins in food products and produce a modifiable structure. For example, they respond to the gluten present in wheat and increase protein elasticity, thereby increasing bakery products volume.
- Adhesion correction: Some emulsifiers are added to food products containing sugar crystals that are scattered in fat and, by coating on glucose crystals, reduce adhesion. This feature affects the fluidity of molten chips and prevent fat appearance on the surface of chocolate.
- Creating foam: Emulsifiers with saturated fatty acids stabilize the bottommost surface of aqueous solutions. Therefore, it is an important factor in creating foam in raw instant desserts.
- Tissue modification is a complex process that is performed on starch and reduces breakdown. For example, homogeneity usually occurs in pasta, bread and bakery products.
- Modifying the dispersion of liquids in another liquid in order to formulate clear solutions: Many greed and colors require emulsifiers for solving.

Today, due to emulsifier effect on human health and limited resources as well as expensivity, researchers have produced emulsifiers using natural resources, especially microorganisms. According to findings, many microorganisms are able to produce compounds with emulsifying properties. A number of these bioemulsifiers have been licensed by the International Organization for Animal Health, including WHO; but most of these compounds have been studied from a nutritional point of view. A large number of biomolecules are also used in the oil, food, drug and chemical industries. The schematic and mechanism of action of important emulsifiers produced by microorganisms using biotechnological processes are presented in Figure.

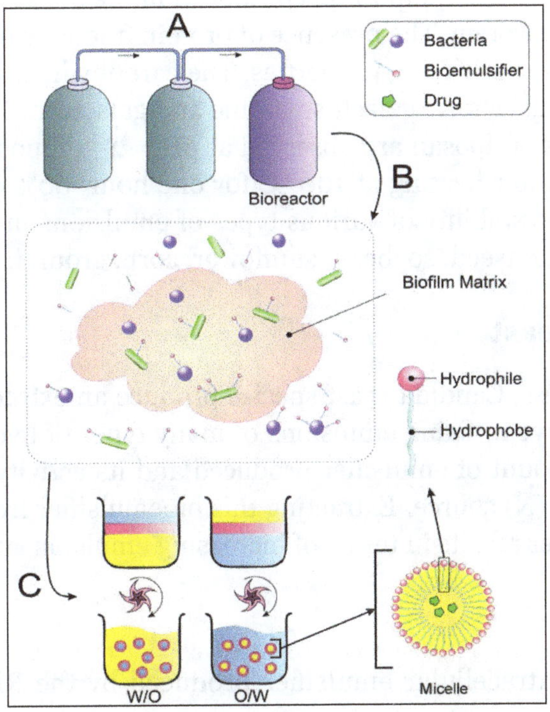

The schematic and mechanism of action of bioemulsifiers in emulsion systems. A: The cultivation, preparation, purification and isolation of microorganisms producing bioemulsifiers. B: Bioemulsifiers produced by microorganisms. C: Emulsion production, adding bioemulsifiers to emulsion systems and assessing their stability.

Bioemulsifiers Derived from Yeast and Fungi

Mannoprotein

Mannoprotein bioemulsifier is a glycoprotein with a molecular weight of about 14,000 to 15,800 Dalton. Within the cellular wall of Saccharomyces spp. and Kluyveromyces marxianus of yeast, mannoprotein molecules are present in glucan, networks, and released from the cell wall of yeast using pressurized heat treatments. This bioemulsifier is able to stabilize oil-in-water emulsions (O/W). According to researchers, bioemulsifiers can be used for producing mayonnaise along with carboxymethyl cellulose (CMC), instead of using expensive ingredients such as ginseng for mayonnaise formulation. Bakery's yeast (Saccharomyces cerevisiae) is an affordable, inexpensive and non-toxic source used for producing this bioemulsifier. Mannoprotein is stable in pH = 3-11. Removing mannoprotein molecules from bakery yeast cell is possible using thermal and enzymatic processes (β-1 and 3 glucanases). Mannoprotein molecules are formed from a polypeptide chain with short and long mannose links. When the protein portion of the mannoprotein molecule is detached by the protease enzyme, the mannoprotein emulsifier disappears. In an industrial scale, this bioemulsifier is active at concentrations equal to or greater than 5% sodium chloride.

Liposan

Liposan bioemulsifier is a water-soluble emulsifier obtained from extracting organic solvents fermented by Candida lipolytica yeast. Liposan is produced in the extracellular layer and consists of 83% carbohydrate and 17% protein. The presence of protein fractions in the bioemulsifier polymer molecule is essential for its emulsifying properties. The carbohydrate part is a heteropolysaccharide, composed of glucose, galactose, galactose-amine and galacturonic acid molecules. The maximum liposomal properties of liposan are observed at pH = 2-5. Liposan is resistant to temperatures less than 70 °C, but after heating at 100 °C for one hour, 60% of its emulsifiable strength reduces. Liposan causes the stability of various types of emulsions in oil, such as hydrocarbons, vegetable oils including cottonseed, soybean, sunflower, corn, ground, safflower and olive oil.

Candida Tropicalis Yeast

During the fed-batch process, Candida yeast species produce an extracellular bioemulsifier. This bioemulsifier is very effective in fixing emulsions of many types of hydrocarbons, especially aromatic compounds. The amount of emulsifier produced and its activity increases during fermentation by limiting nitrogen (N) source. Extracting this bioemulsifier from Candida tropicalis cells using hot water shows better results in terms of increasing emulsion strength.

Rhodotorula Yeast

This bioemulsifier is an extracellular emulsifier produced by the Rhodotorula glutinis yeast.

It is formulated during fed-batch fermentation and glucose utilization under limited nitrogen conditions at 30 °C and pH = 4.

Phaffia Yeast

Phaffia rhodozyma is a basidiomycetous pink yeast. It has been known as a natural source of astaxanthin and many other nutrients. Also, it is currently being used as an ingredient in feeds. It grows on carbohydrate, hydrocarbon and a mixture of carbohydrate and lipid polymers. Fermentation is accomplished in a culture medium chain with sucrose as the carbon source for 3 days at 22 °C and centrifuged at 150 rpm. Experiments have shown that adding sodium citrate, stimulates bioemulsifiers production.

Table: List of bioemulsifiers producing yeast.

Microorganism's	Bioemulsifiers
Torulopsis petrophilum	Sophorolipids
Torulopsis apicola	Sophorolipids
Pseudozyma rugulosa	Mannosylerythritol lipids
Pseudozyma aphidis	Mannosylerythritol lipids
Kurtzmanomyces sp.	Mannosylerythritol lipids
Kurtzmanomyces sp. I-11	Mannosylerythritol lipids
Debaryomyces polymorphus	Carbohydrate protein-lipid complex
Saccharomyces cerevisiae	Mannoprotein
Kluyveromyces marxianus	Mannoprotein
Rhodotorula glutinis	Polymeric bioemulsifier

Table: List of bioemulsifiers producing fungi.

Microorganism's	Bioemulsifiers
Candida tropicalis	Mannan-fatty acid
Candida lipolytica Y-917	Sophorous lipid
Candida utilis	NDA
Candida ingens	Fatty acids
Candida lipolytica UCP0988	Carbohydrate-protein-lipid complex
Candida tropicalis	Liposan
Candida bombicola	Sophorolipids
Candida (torulopsis) apicola	Sophorolipids
Candida lipolytica ATCC 8662	Carbohydrate-protein complex
Penicillium chrysogenum	Polyketide derivative
Yarrowia lipolytica IMUFRJ 50682	Carbohydrate-protein complex
Yarrowia lipolytica NCIM 3589	Bioemulsifier

Yarrowia lipolytica IMUFRJ 50682	Yansan
Ustilago maydis	Cellobiose lipids
Candida sphaerica UCP0995	Sophorolipids
Candida. glabrata UCP0995	Sophorolipids
Pseudomonas. aeruginosa	Rhamnolipids

Bioemulsifiers Derived from Bacteria

Lauryl Fructose

This bioemulsifier is produced by the lipase enzyme obtained from Pseudomonas spp. in a culture media containing dry pyridine. This bioemulsifier has emulsification properties for a variety of hydrocarbons, edible oils, and oil based oils such as margarine and shortening. In a water-containing environment, Laura fructose reduces surface tension from 72 to 29 (MN/m). Also, it reduces the intermolecular reaction of water and hydrocarbons from 50 to 6 when combined with water-insoluble oil compounds.

Table: List of bioemulsifiers producing bacteria.

Microorganism's	Bioemulsifiers
Pseudomonas fluorescens	Viscosin
Pseudomonas aeruginosa	Rhamnolipids
Pseudomonas fluorescens	Carbohydrate-lipid complex
Bacillus amyloliquefaciens,	Surfactin/Iturin
Bacillus subtilis	Subtilisin
Bacillus subtilis	Lichenysin
Bacillus licheniformis K51, Bacillus subtilis	Peptide lipids
Bacillus pumilus A1	Rhamnolipids
Bacillus sp. AB-2	Hydrocarbon-lipid-protein
Acinetobacter calcoaceticus	Emulsan
Acinetobacter radioresistens	Alasan
Acinetobacter calcoaceticus RAG1	Emulsan
Rhodococcus erythropolis	Glycoprotein
Rhodococcus sp. 33	Uronic acids
Cyanobacteria	Trehalose dicorynomycolate
Clostridium pasteurianum	Polysaccharide
Debaryomyces polmorphus	Whole cell
Halomonas eurihalina	Carbohydrate-lipid complex
Halomonas	Emulsifier HE39 & HE67
Lactobacillus paracasei	Glycoprotein
Leuconostoc mesenteriods	Dextran
Serratia marcescens, Serrated rubidea	Serrawettin

Bacillus pseudomycoides BS6	Lipopeptide
Pseudomonas. cepacia CCT6659	Rhamnolipids
Bacillus. licheniformis R2	Lipopeptide

Alasan or E-KA53

Alasan is a biomolecular bacterium produced by the Acinetobacter radioresistant bacteria. The molecular composition of this complex bioemulsifier consists of polysaccharides and proteins with high molecular weight (100,000 - 230,000 Daltons). If the protein portion is damaged and digested using proteolytic enzymes, the bioemulsifier polymer turns into a thick polysaccharide and loses its emulsifying properties. Heating with alcohol at 50 °C may lead to 2.5 times increase in polysaccharide concentration, while the protein portion and the emulsifying properties of the molecule remain unchanged. Heating at 60-90 °C reduces viscosity and increases emulsifying activity as much as 5.8 times than the initial value. The emulsifying properties of alasan are affected by pH and magnesium ion concentrations. This bioemulsifier is an extracellular product and is used extensively in the food industry. Alasan is produced during Acinetobacter bacterium fermentation in a fed-batch fermentation system. The ultimate product consists of 2.2 g of emulsifier per liter of culture fluid. Alcohols stabilize a wide range of oil-in-water (O/W) emulsions, such as n-alkanes, alkaline compounds, liquid paraffin, soybean oil, coconut oil and raw oils. The emulsifying activity of alasan increase approximately 2-3 times, when heated at 100 °C under neutral or alkaline conditions. This bioemulsifier is completely active in pH = 3.3-9.2 and its maximum emulsifier activity is at pH = 5. Magnesium ions increase the activity of emulsifiers both at lower (3.3-4.5) and higher pH (5.5-9.3) than that of the optimal pH. Alasan activity is higher in an environment containing 20 mL citrate than that of same concentrations of acetate or Tris-HCl. According to studies, we can indicate that this bioemulsifier is a high molecular weight anion heteropolysaccharide combined with a protein component such as alanine.

Emulsan

Emulsan is an extracellular poly-anionic bioemulsifier produced by Acinetobacter calcoaceticus RAG 1 bacteria. In fact, emulsan is a lipoheteropolysaccharide polymer containing D-galactose-amine produced during the stationary phase. This bioemulsifier is a poly-anionic and amphiphilic compound which is able to stabilize the hydrocarbon emulsion in water by creating a very thin layer between the hydrocarbon droplets and water. Maximum concentration is obtained when culture media containing 12 carbon-based fatty acids are used as the carbon source. Emulsan production is possible with fermentation methods such as batch, chemo-stat, immobilized cell system and self-cycling fermentation (SCF). Based on SCF methods, bioemulsifier production could increase about 50 times compared to the batch method. Another type of emulsan considered as bio-emulsion is produced by Acinetobacter calcoaceticus, which is used in the formulation and production of soft cheese and ice creams as well as creams and skin-protecting materials. Different types of emulsions produced from these bacteria include alpha-amyloemulsan, Apo-alpha-oleo-emulsan, and beta-emulsan, which are used for treating skin infections are widely used in the food industry. These compounds are mostly poly-anionic lipoheteropolysaccharides that are produced by different species of Acinetobacter venetianus rag-1t ATCC 31012- and Streptomyces NRRL.B-15615, NRRL.B-15847 or ATCC 31926 and few other species.

Cyanobacteria

A variety of cyanobacteria (Genus phormidium, ATCC 39161) (Oscillatoriales) bacterium produce bioemulsifiers that can be used for producing hydrocarbon and oil emulsions in a fluid environment such as water. This bacterium is obtained from using precise separation methods from riverside water, which subsequently grow on a suitable culture medium under favorable conditions and produce an extracellular bioemulsifier sphincter. The molecular weight of this polymeric bioemulsifier is more than 200,000 Dalton. According to chemical tests, it contains sugars, fatty acids, and a protein fractions. Also, more accurate tests using IR spectrophotometry have shown that it is contains amide, carboxylic and amino groups. This bioemulsifier is used for producing various types of oil-in-water emulsions (O/W).

Pseudomonas Cepacia Bacteria

This bioemulsifier which in terms of molecular characteristics, is considered as a mixture of glycolic acid, is produced after Pseudomonas cepacia bacteria growth and propagation on sunflower oil (as the carbon source) medium. The production of this bioemulsifier is carried out by Pseudomonas cepacia bacteria by adding 1.7 % of sunflower oil per liter of culture medium when oxygen and nitrogen levels are controlled. This bioemulsifier is used as a natural source of disintegrating agents used for decomposing and neutralizing polychlorinated biphenyls, especially polychlorinated biphenyls. Polychlorinated biphenyls are toxic and with carcinogenic compounds produced by pesticides during various chemical processes which are able to contaminate industrial wastewater and cultivated soils.

Bacillus Stearothermophilus

During growth, Bacillus stearothermophilus VR-8 produces an extracellular bioemulsifier on a medium containing 4% crude oil. The optimum temperature for producing this bioemulsifier is 50 °C, which at this temperature, 0.6 gr/L bioemulsifier is produced. This emulsifier is purified by acetone and dialysis and contains 46% protein, 16% carbohydrate and 10% fat. Its emulsifier activity is stable over a wide range of temperatures, i.e. 50-80 °C, and pH = 2-8, and salt concentrations (5% chlorine, 5% calcium chloride or 1% chlorine magnesium). The emulsifying properties of this bioemulsifier are related to its stability in a wide range of pH (liposomal at pH = 2-5 and a maximum of 70 °C) and temperatures. This emulsifier is used for removing crude oil from reservoirs and eliminating the remaining oil of crude oil tanks. Hereby, recovery of crude oil, increases by complete scouring of oil remained on storage chamber.

Sphingomonas Bacteria

Presence of polycyclic aromatic hydrocarbons (PAH) in water resources due to their low solubility, is somewhat problematic. When the molecular weight of these compounds (PAH) are reduced, available microorganisms metabolize them. A number of degrading bacteria have been isolated from multi-ring aromatic hydrocarbons of contaminated soils, which produce bioemulsifiers and active compounds, most notably strain No. 107 of Sphingomonas bacteria. This bacterium grows on culture media containing a variety of aromatic hydrocarbon compounds and create clear spots on the medium. Also in liquid culture media, this bacterium uses aromatic hydrocarbons as the main source of energy and carbon. Notably, this bioemulsifier has emulsifying properties similar to high molecular weighting polycyclic hydrocarbons.

Emulsifying Salts

Dairy products like processed cheese and processed cheese spreads and food products like cheese analogues rely on the action of one important food additive – emulsifying salts (ES). Cheese analogues are substitute of natural cheese that are made using processed cheese manufacturing technique, but utilizing raw materials like rennet casein or acid casein or Na- caseinate or vegetable proteins as protein source, and vegetable fat or even milk fat as the fat source, along with other ingredients (viz., cheese flavor, NaCl, acid, color, preservative, etc.). Each food additive has its own contribution in cheese systems. Among various food additives, ES play a major role in the manufacture of processed cheese, processed cheese spread and cheese analogues. Similar to the proprietary blend of stabilizer-emulsifiers for the ice cream industry, several manufacturers are taking advantage of the proprietary blend of emulsifying salts that are claimed to give excellent functionality to the processed cheese makers.

Understanding of the role of emulsifying salts in such products can help the cheese maker to improve the product quality and may enable the food processors to 'tailor-make' the cheese as per the whims and wishes of the retailers and consumers.

Emulsifying salts (ES), also referred to as 'melting salts' or 'chelating salts' are Na, K or Ca- salts of citrate or phosphates (short or long chained), preferentially used as an ingredient in the manufacture of processed and imitation cheeses. Such additive helps in forming the structure of food and exerts positive influence on the melting properties by aiding emulsification of milk/vegetable fat in the protein matrix. They provide uniform structure during the melting of process cheese and also affect the physical, chemical and microbiological quality of resultant product.

Processed Cheese and Cheese Spreads

The ripened cheeses of various ages are mixed in certain proportion and processed in presence of ES to yield 'processed cheese (PC)/processed cheese spread (PCS)' having enhanced shelf life. The PCS is formulated to have higher milk fat and moisture content so that it has the property to 'spread' during food applications. PC or PCS is manufactured by blending shredded natural cheeses of different types and degrees of maturity with emulsifying salt, adding water, salt, milk fat and heating the blend (80-85 °C/4-8 min) with constant agitation, until a homogeneous mass is obtained. The product is subsequently stored at 5-10 °C after packaging and cooling. The legal requirements (Food Safety and Standards Authority of India, FSSAI) for processed cheese and cheese spread in India specifies use of emulsifying salts such as Na, K, Ca-salts of citric acid, orthophosphoric acid, polyphosphoric acids at a level not exceeding 40,000 ppm (i.e. maximum 4.0% by weight).

In processed cheese industry, a term commonly used to describe the emulsification of cheese and water into a continuous system is 'creaming reaction'. The creaming reaction is related to the protein-protein interactions and the properties of the final product can be related to the level of protein dispersion and reassociation during processing.

Cheese Analogues

'Cheese analogues' also referred to as 'imitation cheeses' are the products made out of dairy, partial dairy or even non-dairy ingredients, that tend to resemble a particular cheese variety. Examples of analogue cheeses are Mozzarella cheese, cream cheese, process cheese, etc. Cheese analogues are prepared by blending dry ingredients (e.g. rennet/acid casein or caseinates, common salt, starch, cheese flavour) in the aqueous solution of ES, followed by heat processing (~80-82 °C/3-5 min) with simultaneous incorporation of vegetable or dairy fat. An appropriate acid (in solution form) is used to adjust the pH of cheese analogue. Further processing is carried out in a manner similar to processed cheese. Imitation cheese products are mainly used in the catering industry and in the processing sector, e.g. in preparation of pizza and prepared dishes.

Need to Emulsify Fat in Cheese Systems

All cheese products are essentially 'oil-in-water' emulsions. Natural cheese is a nearly perfect emulsion stabilized by natural surfactants, the cheese proteins. The principal caseins in cheese (αs_1-, αs_2-, β) have non-polar, lipophilic C-terminal segments, while the N-terminal regions containing calcium phosphate are hydrophilic. This structure allows the casein molecules to function as an emulsifier. However, in absence of ES, heating of natural cheese blend (i.e. of different maturity) above 65°C can result in separation of oil and water phases. To incorporate the fat source (butter/vegetable fat for process cheese and cheese analogue respectively) in the proteinaceous cheese matrix, ES helps in this regard. Emulsification of fat in cheese is affected by factors like amount of calcium in the $CaPO_4$ end, pH, age of cheese and temperature during cheese processing.

In natural cheese, caseins do not possess their emulsifying properties, because they are bound in three-dimensional calcium bond matrix. Therefore, ES are used during processed cheese production. They are capable of exchanging ions and from insoluble calcium para-caseinate more soluble sodium para-caseinate is created, which possesses emulsifying and stabilization function in processed cheese matrix.

Classification of Emulsifying Salts

The ES used for processed cheese systems are mainly classified into following two types: (i) citrates, and (ii) phosphates.

- Citrates: They are salts of citric acid e.g. tri-sodium citrate (TSC) which are preferred in 'block type cheese' and 'sliceable cheeses'. Citrates are highly soluble with fairly good protein dissolving power. They contribute to long, elastic structure to processed cheese; the structure of cheese remains firm and heavy. Few disadvantages of citrates are lack of creaming, lack of bacteriostatic action, and risk of sandiness and mottling in cheese.
- Phosphates: They are salts of phosphoric acid. There are two types of phosphates i.e. (i) Mono phosphates (orthophosphates) e.g. NaH_2PO_4, Na_2HPO_4 and Na_3PO_4 and (ii) Condensed phosphates e.g. $Na_6P_4O_{13}$; $(NaPO_3)_n$.
    - Monophosphate/orthophosphates: These include mono-sodium, di-sodium and tri-sodium orthophosphates (TSP). Mono-phosphates are water-soluble and have excellent

buffering properties, but do not exert any creaming effect. Mono-sodium and TSP are most commonly used for pH adjustment.

- Condensed phosphates: These are classified into chain forming polyphosphates, ring-forming metaphosphates and cross-linked ultraphosphates. The chain forming polyphosphates are classified as short and long-chain polyphosphates. Polyphosphates have strong creaming action and exert bacteriostatic property. Their buffering capacity depends on their chain length.
- Short chain polyphosphates: These are mainly the diphosphates and tripolyphosphates with pH value of 1.0% solution ranging from 2.7 to 10.2. The commonly used salts are mono-sodium, di- sodium, tri-sodium and tetra-sodium diphosphates.
- Long chain polyphosphates: They have an average degree of condensation ranging from 4 to 25. Graham's salt (i.e. Sodium hexa Meta Phosphate – SHMP) with a degree of condensation of 10-25 has outstanding ion-exchange ability. The disadvantage is their proneness to partial hydrolysis during processing.

Role of ES in Cheese System

In order to modify the cheese proteins in natural cheese so that they can emulsify the fat during processing and to develop certain characteristics in processed cheeses like uniform body, melting of cheese during its application, ES are made use of.

The mechanism of action of emulsifying salts in any cheese system involves the following:

- Removal of calcium from the protein system.
- Peptizing, solubilizing and dispersing the proteins.
- Hydrating and swelling of the proteins.
- Emulsifying the fat and stabilizing the emulsion.
- Controlling pH.
- Forming an appropriate structure of the product during or after cooling.

The monovalent Na and bivalent Ca ions are antagonistic with regard to their action on proteins. The calcium bound to casein fragments depresses their water solubility. ES removes calcium by exchanging it for sodium or by binding to it. The functional behavior of cheese products are strongly influenced by the type of ES and its physico-chemical properties, including its ability to bind calcium, casein dispersion during cooking, and creation of cross-links with casein during cooling.

Type of Emulsifying Salts

The most commonly used ES for process cheese manufacture are TSC and disodium phosphate (DSP). TSC is the preferred ES for 'Slice-on-slice' process cheese varieties, whereas DSP (or appropriate combinations of di- and trisodium phosphates) is used in 'Loaf-type process cheese' and 'Process cheese spreads'. Sometimes, low levels of SHMP are also used along with these ES.

The quality of ES added to processed cheese blend depends not only on the type and age of cheese used in the blend (proportion of water and calcium), but is also determined by the final product group (i.e. block, spread, slices, etc.). The ES that help in structure building (not creaming) include high molecular polyphosphate like C, SE, S7, PZ (ES of Joha company) and citrate, which leads to firm, slicing processed cheese. ES that lead to creaming, e.g., lower and medium molecular polyphosphate, S9, S9 special, S10, S90 (ES of Joha company) are suitable for spreadable processed cheese.

The ability to exchange ions and form three-dimensional structure during cooling is different for individual ES and decreases in the order: polyphosphates > triphosphates > diphosphates > monophosphates ≈ citrates. Citrates have higher ability to exchange ions than monophosphates. Longer phosphates lead to a higher dispersion of casein due to more intensive ability of ion exchange.

A combination of Joha S9S + Joha NO (1:1) ES @ 3.0% w/w was found suitable to produce desired PCS containing whey protein concentrate (WPC).

Rate of Addition of an Emulsifying Salt

The amount of ES that can be added to the cheese base are regulated by many countries and usually do not exceed 3.0 or 4.0%. Increased levels of polyphosphate in ES mixtures (with pyrophosphate or orthophosphate) resulted in a firm PCS based on Ras cheese, whereas increased levels of pyrophosphate produced softer cheeses; buffering capacity increased with increasing levels of orthophosphate and pyrophosphate. The degree of casein dissociation, pH and hardness of PC increased as the concentration of ES was raised. The PC prepared with TSC was whiter than those prepared with other emulsifying salts.

The hardness of non-fat Pasta filata cheese decreased, while the meltability increased with increasing concentrations of TSC (1.0 to 5.0%). The effect was attributed to a decrease in the number of colloidal calcium phosphate cross-links and an increase in electrostatic repulsion in the cheese system. Addition of 1.0% TSC increased the stretchability of cheese, by partially loosening the protein matrix.

The concentration of the ES affected the sensory properties of the cheese product. The meltability of processed cheese increased with an increase in the ES content. The hardness of processed cheese decreased as the ES content was raised.

Combining Emulsifying Salts

A single ES is not capable of providing all of the desirable characteristics to process cheese or cheese analogues. The right proportion of individual ES should be arrived at and adopted. TSC and DSP when added in 1:2 ratio yielded an excellent quality processed cheese with desired firmness, sliceability and smoothness. Such combination helped to strike a balance between the water absorption properties of diphosphate, and the protein solubilizing property of citrates.

Processed cheese produced using mixtures of sodium diphosphate, sodium polyphosphate and sodium tripolyphosphate (STPP) (30:40:30) yielded better quality product than that made using Joha salts SE and PZO.

Impact of Emulsifying Salts on Properties of Cheese

Inclusion of an ES has influence on milk proteins and hence its interaction with milk fat in cheese system; variation in the type and/or amount of ES is bound to affect the cheese properties. The influence of ES on some important properties of cheese system is dealt hereunder.

- Appearance: Increasing the concentration of ES tended to improve the whiteness of processed cheese spreads based on Ras and Quark cheeses.
- pH and pH stability: Applying an appropriate mixture of ES causes an increase of the cheese blend pH from 5.0–5.5 up to 5.6–5.9 and, furthermore, contributes to pH stabilization due to its buffering capacity. Due to this particular pH increase, the calcium-masking ability of the ES is increased by the negative charges on the para-caseinate, thus promoting a concomitant increase in hydration and solubility of para-casein and the formation of a stable product.
- Textural properties: The influence of phosphate type emulsifying salts on cheese firmness increased in the order: orthophosphate, polyphosphate, diphosphate and triphosphate. TSC and DSP exhibited similar behaviour and produced the softest, most adhesive, least elastic and most liquid-like final processed cheese products that showed reduced apparent viscosity during processing. TSPP gave harder, less adhesive, more elastic and more solid-like samples compared to use of penta-Na-tripolyphosphate.
- Melt, flow properties and sliceability: The variation in the TSC concentration (2.0-3.0%) influenced only the melt and flow characteristics, but did not affect the unmelted textural properties of processed cheese. TSC contributed to greater 'meltability' and 'sliceability' to process cheeses compared to several other emulsifying salts; the process cheese had a firm, gel- like structure. TSC in combination with DSP @ 3.0% level produced significantly softer cheeses and melted more easily than those prepared with condensed phosphate Joha salts. Meltability of cheese was in decreasing order for ES: trisodium pyrophosphate > DSP > TSC. TSP and DSP were reported to produce the most pronounced melting effects. Sodium citrate induced significant melting only at the highest tested concentration (3.0%).

Bacteriostatic Action of Emulsifying Salts

ES such as orthophosphate and polyphosphate may inhibit growth and toxin production by Clostridium botulinum by sequestering Fe, Mg or Ca ions. Polyphosphates may also interact physically with bacterial cells by forming channels, increasing their permeability to inhibitory compounds, and promote leakage and cell lysis. Growth and gas formation by Cl. tyrobutyricum were prevented by addition of 0.5 - 1.0% polyphosphate to process cheese. Use of 0.5% long chain polyphosphate formulation (JOHA HBS-1, HBS-9) may be sufficient to control Cl. tyrobutyricum growth under normal conditions, where initial spore counts are low and storage temperature is 20 °C.

Reduced Sodium Processed Cheese

The average sodium content was reported to be 402 mg, 335 mg and 382 mg per 28 g each of processed cheeses, cheese food, and cheese spread respectively. The proposed sodium labeling for foods combined with a recommended dietary goal of substantially reducing sodium intake in several countries has focused attention on technological means to reduce sodium in processed foods. ES such as potassium citrate, potassium acetate, and mixtures of these at ratios (i.e. 1:1, 2:1 and 1:2), @ 3.0% by weight have been used in order to reduce the sodium content of cheese. Defects related to emulsifying salts. If the ES is not utilized judiciously in cheese formulation, it may lead to some defects. Fish-eye defect: 'Fish-eyes' (hard, glassy lumps of protein) may be formed in processed cheese due to poor hydration of the protein.

- Bitterness in cheese: An overdose of specific emulsifying agents (i.e. high in phosphorus content) can lead to bitterness in processed cheese slices. Such bitter slices showed very weak αs1- and β-caseins region, had only γ-casein and low-molecular weight peptides, and showed high concentrations of hydrophilic and hydrophobic peptides.
- Crystals: Crystals of ES (i.e. calcium phosphate/diphosphate, calcium lactate/citrate) may be observed in processed cheese as a result of: (i) Excess use of emulsifying salt, (ii) An unsuitable mixture of emulsifying salts, or (iii) Insufficient dissolution of the latter in the mixture during processing.

Emulsifying Salts for Cheese Analogues

When using rennet casein or caseinates, they must first be solubilized by the addition of emulsifying salts, in the manufacture of cheese analogues. The hot solution of calcium chelating salts (i.e. ES) are used to facilitate hydration of the protein which subsequently acts as an emulsifier for the oil phase dispersed in the protein/aqueous phase of the cheese analogue. The rheological properties of the rennet casein during hydration are dependent on the concentration of ES in the hydrating solution. In the manufacture of rennet casein based cheese analogues, ES are used to disrupt the calcium mediated protein-protein cross-bridges, thereby allowing the protein to hydrate. Such improved hydration enhances the emulsifying ability of rennet casein in cheese analogues. In the formulation of cheese analogue, the type of ES used is dictated by the protein source used (i.e. rennet casein/acid casein/caseinates); the level of incorporation of ES decides the functionality and sensory properties of resultant cheese analogues.

Sodium aluminum phosphate is commonly used in rennet casein/Na-caseinate based mozzarella type imitation process cheese since it provides desirable functional properties for pizza applications. TSP and DSP were found to exert most pronounced melting effects in such systems.

Orthophosphates, such as DSP and TSP, and polyphosphates, such as tetra sodium pyro and SHMP, are approved for cheese use. For standardized cheese products, 3.0% phosphate is allowed, though usually 1.8 to 2.5% phosphate is sufficient. Beyond this level, phosphate can cause crystals to appear in the cheese. If crystals are a problem, sodium aluminum phosphate can be used to avoid them.

Other Applications of Emulsifying Salts

Cheese powder: Cheese powders are used in the food industry as natural functional and flavour ingredients in applications such as biscuits, savoury snacks, bakery products, sauces, dressings, ready meals and processed cheese, typically added at levels of 2-12%. In cheese powder production, cheese is comminuted and melted with addition of water and ES. This cheese slurry, cheese feed, is then heat-treated and finally most of the water is removed by spray drying. ES, primarily sodium phosphate, are added to create an emulsion that remains stable until spray drying and also to ensure a final powder with good keeping quality.

The raw material for cheese powder manufacture is usually a carefully selected blend of cheeses, of different ages. The cheese is comminuted and mixed with water to produce slurry containing 35-45% solids. Stabilizing salts (phosphates, citrates) are added in small amounts (i.e. 2%), to modify the physical characteristics of milk protein so that the fat is effectively emulsified and the slurry melts evenly on heating to give a smooth body. The slurry is pasteurized and homogenized (2000 + 500 psi) before spray drying to produce a powder containing < 3% moisture.

Other Applications in Dairy Industry

Emulsifying salts are used as agents for preventing undesirable changes of ultra-high temperature (UHT)-treated milk, UHT-treated cream, coffee cream, concentrated milk, etc. DSP can be used to reduce UHT milk protein coagulation. Long chain SHMP inhibits UHT milk age gelation. UHT milk, cream and coffee creamers gain stability and shelf life by treatment with JOHA specialties. Such ES contribute to consistency and texture while reducing heat-induced protein coagulation. In non-dairy coffee whiteners, a phosphate buffering system contributes to stability and prevents 'feathering' and 'fat separation', when the coffee whitener is added to the hot acidic coffee medium.

DSP is added to fluid milk prior to pasteurization or spray drying to inhibit protein denaturation during heat treatment and to allow efficient protein dispersion upon rehydration. The purple discolouration of strawberry flavoured milk may be inhibited by addition of sodium tripolyphosphate (STPP) to bind the iron. Chocolate milk may benefit from the presence of trisodium polyphosphate (TSPP), which keeps the cocoa in suspension.

In dairy preparations using alginate as a texturizer, phosphates are ideal for controlling calcium release and consequently, gel setting. DSP, TSPP or SHMP may be added to ice cream to prevent churning of the milk fat.

Sodium citrate is also an important stabilizer in whipping cream and vegetable based dairy substitutes. Non-dairy creamers frequently use dipotassium phosphate (DKP) to inhibit feathering, when the whitener is added to the warm, acidic coffee.

Applications other than in Dairy Industry

Sodium and/or potassium salts of phosphates and/or polyphosphates are regularly used in the food industry, namely meat and bakery industries.

- Meat, poultry and sea foods: Phosphates play an important role in pH adjustment in meat, poultry and sea foods. Alkaline phosphates increase meat pH approximately 0.1-0.6 units, depending upon the particular phosphate used. Processors add phosphates to meat, poultry and fish to achieve moisture retention, along with higher yield and flavor protection, especially in processed meats. Phosphates can reduce oxidation, help retain marinade and cook juices, chelate metals, preserve colour and lend freeze-thaw stability. Adding phosphates enhances stability in finely chopped meat systems, such as bologna or hot dogs, by influencing the pH, ionic strength, protein extraction, divalent cation binding and viscosity.

Phosphates help extract proteins from meat, whether chopped or whole muscle. Phosphate type and pH are critical for optimal performance in meats. Pyro is the most effective phosphate for meats; pH choice varies by application. TSPP extracts muscle proteins to a greater degree than other phosphate salts.

- Baked goods: In the bakery industry, baking powder is an appropriate example to demonstrate the versatile use of phosphates in this area. Baked goods are leavened with phosphates that contribute to texture, colour, rise and desirable crumb characteristics. Bakery leavening systems commonly use anhydrous monocalcium phosphate (MCP), dicalcium phosphate dehydrate, sodium aluminium phosphate and sodium acid pyrophosphates. MCP can also act as a dough conditioner. Phosphate based acid reacts with an alkali (sodium bicarbonate), to produce carbon dioxide gas in batters and doughs of baked products. The resulting gas causes the matrix to form an open, honey-comb like structure that greatly influences texture.
- Soft drinks: The soft-drink manufacturers add phosphate to cola drinks to enhance flavor.
- Antibacterial effect: Antibacterial effects of polyphosphates have been observed. This especially applies to the effects on selected Gram-positive microorganisms (some strains of genera Staphylococcus, Bacillus and/or Clostridium), evaluated as contaminants in several foodstuffs.

Food Stabilizers

Stabilizers are an indispensable substance in food items that categorised as food additives. When added to the food items, they smoothen the texture of the food & give a definite body to the food. Food stabilisers are added in relatively small amount which aggravate the effect of emulsifiers. They give a uniform nature to the product & hold the flavoring compounds in dispersion. Perfect binders for the varied conflicting components, food stabilizers are a must.

Stabilisers According to Food Processing Technology

“Stabilisers are substances or chemicals that allow food ingredients, which do not mix well, to remain in a homogenous state after blending.”

Stabilisers According to Food Law (Directive 95/2/EC), UK

“Stabilizers are substances which make it possible to maintain the physico-chemical state of a foodstuff; stabilizers include substances which enable the maintenance of a homogenous dispersion of two or more immiscible substances in a foodstuff and include also substances which stabilize, retain or intensify an existing colour of a foodstuff.”

Food stabilisers are confused with preservatives. But there is a difference between the two. Stabilisers is used to retain the physical characteristics of the food whereas as preservatives prevent the spoilage of the food. A food stabiliser increases the stability and thickness of the food by binding its large molecules.

Purpose of Stabilisers

- They maintain the consistency of the food.
- Food stabilisers do not allow the separation of ingredients that are bound together by emulsifiers.

Common Food Stabilisers

- Alginic acid is a popular food stabiliser that is derived from brown algae is used in ice-cream and syrups that we use on desserts.
- Agar agar is also used as a food stabilisers in the food products like canned meat and pudding.
- Carrageenan is like agar agar that is also obtained from red algae. It is used in shampoos.
- Then there is gelatin, another food stabiliser.
- Pectin or calcium chloride are used as food stabilisers in dairy products.
- Lecithin and mono- and digycerides.

Applications of Stabilisers for Food

Stabilisers for food are used in many kinds of products. The most common ones that uses the food stabilisers are:

- Ice Cream,
- Margarine,
- Low-fat spreads,
- Dairy products,
- Salad dressings,
- Mayonnaise.

Ideal Characteristics of a Food Grade Stabilizer

- Should be nontoxic.
- Should readily disperse in the mix.
- Should not produce excessive viscosity or separation or foam in the mix.
- Should not clog strainers and filters.
- Should not impart off flavor to the product.
- Should be economical.

Types of Stabilizers

There are a variety of hydrocolloids on the market, including those derived from plants or seaweed, and those produced by microorganisms. In general, hydrocolloids have a sugar backbone that contains protruding substituents such as esters, sulfates, or additional sugars. Hydrocolloids available for food applications are either neutral or negatively charged.

Gelatin

Gelatin is a major ingredient in some popular foods, including desserts, candies, stock, consommé, and aspic. It's a versatile substance that has many other uses besides being a food additive. In its pure form, it's a pale yellow, odorless, and almost tasteless substance. Once dry gelatin is mixed with water and other ingredients, it can form some very appealing foods with a pleasant mouthfeel.

Gelatin is produced by boiling collagen obtained from animals. Collagen is the most abundant protein in the animal and human body and is found in connective tissue, which connects and supports other tissues. The collagen that's used for gelatin production is generally obtained from pig skin, cow skin, or cow bones, but it's sometimes obtained from the skin of fish instead. It isn't made from horns or from horse hooves, as is sometimes thought.

Gelatin is sold in a powder form or as sheets called leaves. When solid gelatin is added to hot water, a liquid mixture known as a colloid forms. This mixture thickens as it cools and eventually turns into a gel. Gelatin has many useful functions beyond the kitchen. It's used to forming gel caps for medications, for example. It's also used in a variety of cosmetics, hair treatments, and crafts.

Colored and flavored jelly powder

A Problem in Special Diets

Gelatin is widely used in the food industry, but its addition to food creates problems for certain people. Some people object to the use of gelatin for religious or ethical reasons because it's produced from animal tissues or because it's made from the tissues of specific animals.

Any product made from pig tissue is unacceptable in some religions. People with certain religious beliefs require animals to be killed in a particular way (ritual slaughter) before they will eat any product obtained from them. Vegetarians eat eggs and dairy but no other food derived from animals. Vegans eat only plants. Both groups avoid gelatin.

Plant-Derived Gelatin

The joy of a jelly or jello dessert

Some plant materials form gels in water and can be substituted for gelatin, but they have slightly different properties from the animal product. Nevertheless, they are a good option for people who don't want to use a substance obtained from animals. Two of these gelatin substitutes are agar (also called agar-agar) and carrageenan.

How to Make Gelatin

Gelatin can be made at home by boiling animal skin and bones. The boiling process converts the collagen in these materials into gelatin. The process of making gelatin in industry is similar, but additional steps are performed to improve the quality of the substance.

The animal parts are first washed and degreased. They are then treated with an acidic or basic solution to remove minerals and to make the collagen easier to access. Next, the processed animal parts are boiled in distilled water to produce gelatin. Finally, the water is evaporated from the liquid and the solid gelatin is pressed into sheets or ground into a powder.

Composition of Gelatin

Gelatin is a mixture of proteins and peptides obtained from the partial breakdown of collagen. A protein is a long chain of amino acids; a peptide is a shorter chain of amino acids. Gelatin is rich in glycine and proline, two amino acids that are said to be "non-essential" because our bodies can make them. On the other hand, it doesn't contain tryptophan and is very low in threonine, isoleucine, and methionine. These four amino acids are said to be essential because we can't make them and must obtain them pre-formed in our diet. Gelatin is a popular and useful food additive, but it's not a very nutritious source of protein for humans because it lacks some of the essential amino acids.

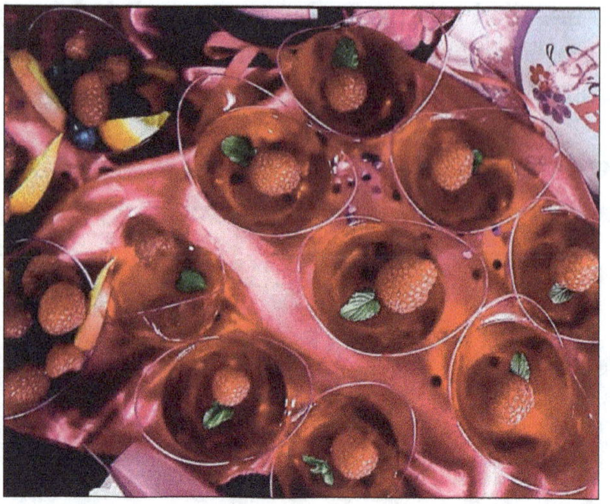

Gelatin and fruit

Properties and Behavior

Gelatin in water is a hydrophilic colloid. A colloid is a mixture in which the particles of a substance are dispersed through a continuous medium. The particles are bigger than those in a solution but are still small. They aren't visible to us, can't be filtered, and don't settle to the bottom of the container. The word "hydrophilic" means that gelatin attracts water.

When gelatin is added to hot water, the bonds holding amino acid chains together break. The chains then move through the liquid, becoming tangled. As the water is cooled, new bonds form between the tangled chains. In some places the chains form a triple helix (three amino acid chains joined together and twisted into a spiral). Water is trapped in the spaces within the helices and tangles, forming a gel.

Uses of Gelatin in Foods

The gel formed by gelatin in water is stable over a specific temperature range (which depends on the quality of the gelatin and its concentration). It breaks down when it's heated beyond the highest temperature in this range. The gels that are added to foods such as marshmallows break down at below body temperature. This means that when the food is eaten, it softens and collapses, creating a "melt in the mouth" sensation and releasing flavor trapped in the gel.

Gelatin is used to make gelatinous desserts, gummy candies, and many yogurts as well as some marshmallows. It's also added to certain meats to prevent them from drying out and to give them an attractive glaze. Stocks and consommés form gels as they cool. In canned meats, gelatin absorbs juices released from the meats as they are processed under pressure. In pâtés, the substance helps to stabilize the emulsified fat. Gelatin is useful for its thickening, emulsifying, binding, and adhesive properties. In addition, it's able to attract impurities and clarify fruit juices, wines, and vinegar.

Stock and Consommé

Stock is a liquid produced by simmering meat and vegetables in water for a long time. The solids provide flavor to the stock and the meat provides collagen. After hours of cooking, the solids are filtered out to leave the liquid stock.

Consommé is clarified stock. Egg whites are added to the stock and heated. As the egg whites solidify, they absorb particles from the liquid and form a raft on its surface. A clear consommé is located under the raft.

Aspic

An aspic is a savory dish in which a gelatin gel made from a meat stock—or more commonly from a consommé—surrounds and encases meat. The meat stock is made by simmering animal bones, meat scraps, and vegetables in water. These may all be roasted first to intensify their flavor. After the mixture has been simmered for a long time, the solids are filtered out to leave the liquid stock. The stock is cooled slightly so that it thickens and is then poured over meat in a mold. The stock may need added gelatin in order to gel, depending on the amount of bone that was added to the pot.

A colorless gelatin gel is often used in commercial aspics so that the meat inside is clearly visible. The gel can make it easier to cut the meat without causing it to disintegrate. The word "aspic" also refers to a dish in which a gel encases vegetables.

Gelatin in Desserts

Gelatin is often added to low-fat dairy products such as yogurt, ice cream, and buttermilk in order to provide the sensation of eating fat. It may be added to some full-fat products as well. The gelatin provides a smooth and creamy mouth-feel and the breakup of the gel in the mouth resembles fat melting. Gelatin in creams and whipped toppings helps to stabilize their consistencies.

Fresh pineapple and papaya contain enzymes that stop a gelatin gel from forming. The chemicals prevent the formation of the bonds needed to hold the amino acid chains together as the gel forms. The enzyme mixture in pineapple is called bromelain. The enzyme in papaya is papain. Canned pineapple doesn't contain bromelain since the heat applied during the canning process destroys the enzymes.

Other foods that reportedly prevent the gel from forming include kiwi, figs, guava, and ginger root. Any form of these foods that has been heated during pasteurization or another process should allow the gelatin to set, since the critical enzyme or enzymes will have been destroyed.

Agar

Agar agar, or simply called agar is a dried, hydrophilic, colloidal polysaccharide extracted from one of a number of related species of red algae (class Rhodophyceae).

Uses

The main purpose of using agar in food with its following three advantages:

- A natural ingredient, suitable for vegetarians to replace gelatine derived from animal skin and bones.

- The high gelling hysteresis between the gelling and melting temperatures.
- Gel reversibility.

What is Agar Agar Made of?

It is a polysaccharide, consisting primarily of D- and L-galactose units. About every tenth D-galactopyranose unit contains a sulfate ester group. Calcium, magnesium, potassium or sodium cations are also associated with the polysaccharide.

It is a complex mixture of polysaccharides, not only consists of two major polysaccharides, agarose and agaropectin, but also with other varieties, some are rich in sulfate, others in pyruvate. Its composition depends on the sources of seaweed and the production process employed.

Agarose

The major component of agar, around 70% or more of its total weight. It provides the gelling power and is free of sulfates. Agarose is made up of the repeating unit of agarobiose, which is a disaccharide, composed of D-galactose and 3–6,anhydro- L-galactose, contributing to the main structure of agar.

The following is the agarobiose structure:

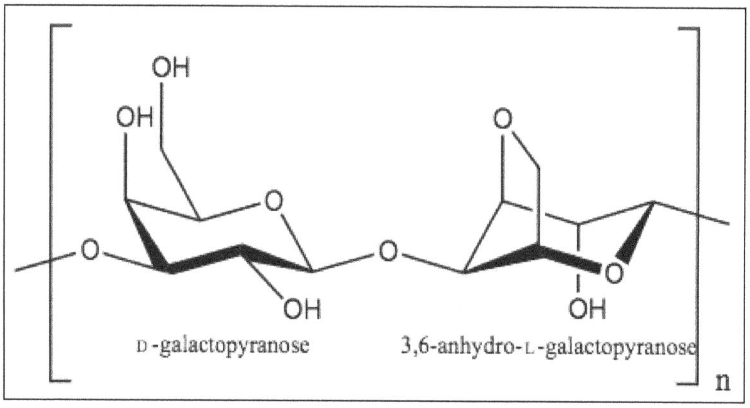

Agaropectin

The non-gelling fraction, a charged sulfated polysaccharide that will influence solution properties, gel strength and gel features.

How is Agar Agar Made?

In general, the manufacturing process of agar powder has the following steps:

- Algae harvesting: Agar plants mature in summer.
- Preliminary drying: The moisture content can be reduced around 80% by natural or artificial dehydration.
- Bleaching: Sodium hypochlorite and/or H2O2 are used.

- Extraction: After this process, agar is dissolved in boiling water. This process may be different depending on the type of agarophyte and the product quality needed. Gelidium is processed under pressure (105–110 °C for 2–4 h) and Gracilaria is usually treated with water at 95–100 °C for 2–4 h.
    - Before the extraction, alkaline treatment (with sodium hydroxide) is usually employed for Gracilaria to hydrolyze the sulfates and so the agar gel-forming ability increased. Otherwise, the gel strength is too low for commercial use.
- Rest steps: Filtration, gelling and freezing/thawing, drying, milling and packing.

Four Types of Agar Agar

According to the different appearance and application, food grade agar in the market can be classified into 4 types. Respectively bar, strip (string), flake and powder. Agar powder is mostly used for industrial applications. Flakes, bars and strips are mostly used in cooking.

Strips and Bars

Strips and bars are the old forms which are manufactured through a more traditional production method by freezing and thawing the agar gel. Sometimes they are called natural agar. The gels are cut into square bars or extruded to produce spaghetti-like strips 25–40 cm long after cooling down. These two types are produced mostly from Gelidium and commonly used at home to prepare traditional dishes.

Before cooking, these two forms need to soak before soften, and then put in boiled water to make it completely dissolved. Later add other ingredients, such as sweeteners, colors, flavours, or fruits to make the recipes.

Flakes and Powder

Flakes and powder are smaller size ones that can quickly dissolve in boiling water. Soaking may be needed for flakes before cooking. The powdered form is the most used one.

Specification

Other names	Agar, Gelose, Kanten, Ceylon, Chinese or Japanese isinglass
CAS number	9002-18-0
Chemical formula	NA
Molecular weight	NA
Appearance	Agar powder is white to yellowish-white or pale yellow, odorless or has a slight characteristic odour.

Properties

Solubility: Insoluble in cold water; soluble in boiling water.

Gelation: Gel strength comes from the composition of agarose.

How to Form a Gel?

Heat is required to make agar completely soluble and the solution will gel upon cooling. Unlike carrageenan, it does not require the presence of any particular ions to gel. Also, it is different with high methoxyl pectin, the gel of which depending on sugar.

Low Concentration is Needed

It absorbs 20 times of water and swells in cold water. When dissolved in hot water and upon cool, it solidifies at a very low concentration (0.5%). But it cannot be gelled and will become a viscous liquid if the concentration is below 0.1%.

Setting and Melting Point

1% agar solution solidifies at 32 ~ 42 °C, its gel has elasticity, and its melting point is 80 ~ 96 °C. Many of its uses take advantage of this large difference between gelling and melting temperature. When compared with gelatin gels, it is tasteless, odorless, and sets more firmly, even at room temperature. Gelatin gels melt around 35-40 °C.

Reversibility

Agarose produces 'physical gels' which means the gel formed only by hydrogen bonds which make the gel reversible. That's to say, it melts just by heating but gels again upon cooling. These transformations can be repeated. The reversibility is an important property of agar gel. Gelatin also has the same characteristics. This advantage can be used in a lot of applications, such as in food, microbiology, biochemistry and so on. On the contrary, the 'chemical gels' is taken place by covalent bonds which can be considered through a chemical reaction and shape the gel irreversible.

Synergies

Agar obtained from Gelidium is synergistic with locust bean gum (LBG) due to the agaropectin in Gelidium. The increased gel strength, decreased rigidity and enhanced elasticity are observed in the gel formed by the mixture of LBG and Gelidium agar. Agar comes from Gracilaria shows synergy with sugar, resulting in increased gel strength when used in high sugar concentration (around 60%) products, such as jams and jellies.

Uses of Agar Agar

This hydrocolloid is primarily used for its gelling properties. About 90% in the market is used in food applications, with the remaining 10% is intended for bacteriological and other biotechnology applications.

Food

Generally, agar is used in jelly, bakery, confectionery, dairy products, beverage, meat products and so on.

- Jelly: The most use of agar is in jellies and other desserts that need to gel. In Asian countries, agar is a traditional and popular component of jellies. For producing jellies and puddings, agar is boiled in water first, making it completely soluble, then add colors, flavors, sweeteners and fruits to the liquid before cooled and finally pour the liquid mix into molds to form a jelly.
- Bakery: In the baking process, it stabilizes and thickens pie fillings, sugar icings, bread dough, cake glazes, coatings and meringues with the property of withstanding high temperature.
- Confectionery: Nowadays, agar from Gracilaria is often used in confectionery with high sugar content, such as fruit candies, gummies, gums, caramels, marshmallows and etc.
- Dairy Products: Agar can be used in dairy products like yogurts, ice cream, cheese, chocolate and etc.
- Beverage: It acts as a clarifying and refining agent for juices, beers and wines.
- Meat Products: Agar is also used to gel meat and fish products, and function as a gelatin substitute for its higher melting temperature and gel strength.

Xanthan

Xanthan gum is a popular food additive that's commonly added to foods as a thickener or stabilizer. It's created when sugar is fermented by a type of bacteria called Xanthomonas campestris. When sugar is fermented, it creates a broth or goo-like substance, which is made solid by adding an alcohol. It is then dried and turned into a powder.

When xanthan gum powder is added to a liquid, it quickly disperses and creates a viscous and stable solution. This makes it a great thickening, suspending and stabilizing agent for many products. It was discovered by scientists in 1963. Since then, it has been well researched and determined safe. Therefore, the FDA has approved it as a food additive and placed no limitations on the amount of xanthan gum a food can contain.

Even though it's made in a lab, it's a soluble fiber. Soluble fibers are carbs that your body cannot breakdown. Instead, they absorb water and turn into a gel-like substance in your digestive system, which slows digestion. Therefore, your body is unable to digest xanthan gum, and it does not provide any calories or nutrients.

Where is Xanthan Gum Found?

Xanthan gum is found in food, personal care and industrial products.

Food Products

Xanthan gum can improve the texture, consistency, flavor, shelf life and appearance of many foods. It also stabilizes foods, helping certain foods withstand different temperatures and pH levels. Additionally, it prevents foods from separating and allows them to flow smoothly out of their containers.

It's used frequently in gluten-free cooking since it can provide the elasticity and fluffiness that gluten gives traditional baked goods. The following are some common foods that contain xanthan gum:

- Salad dressings,
- Bakery products,
- Fruit juices,
- Soups,
- Ice creams,
- Sauces and gravies,
- Syrups,
- Gluten-free products,
- Low-fat foods.

Alginates

Alginic acid, also called algin or alginate and sodium alginate, is an anionic polysaccharide distributed widely in the cell walls of brown algae, where through binding with water it forms a viscous gum. In extracted form it absorbs water quickly; it is capable of absorbing 200-300 times its own weight in water. Its color ranges from white to yellowish-brown. It is sold in filamentous, granular or powdered forms.

Alginates from different species of brown seaweed often have variations in their chemical structure, resulting in different physical properties. For example, some may yield an alginate that gives a strong gel, another a weaker gel; one may readily give a cream/white alginate, another may give that only with difficulty and is best used for technical applications where color does not matter.

When alginate is added to a liquid, it will act as a thickener. In the presence of calcium ions, a mixture containing alginate will form a gel. The calcium ions insert themselves between individual alginate strands and will allow them to interlock and form a gel.

Sodium Alginate Properties

- Texture: Thermo irreversible rigid and brittle gel in the presence of calcium. The higher the concentration of alginate and calcium, the harder the gel produced.
- Appearance: Clear and transparent in food application.
- Temperature (gels and melts): Not affected by temperature. Gel produced is heat stable up to 150 °C (302 °F) but prolonged heat treatment at low or high PH may destabilize the gel.
- Flavor release: As most hard set gels, poor flavor release.
- Mouthfeel: Lingering and sticky.
- Shearing: Breaks gel.
- Syneresis (weeping): Yes, increases with gel strength.
- Freeze/Thaw stable: Yes, in most applications.

Chemical Forms of Alginates

- Sodium alginate is the sodium salt of alginic acid. Its empirical formula is $NaC_6H_7O_6$.
- Potassium alginate is a chemical compound that is the potassium salt of alginic acid. Its empirical chemical formula is $KC_6H_7O_6$.
- Calcium alginate is made from sodium alginate from which the sodium salt has been removed and replaced with calcium, has the chemical formula $C_{12}H_{14}CaO_{12}$.

How Alginates are Produced

The processes for the manufacture of sodium alginate from brown seaweeds fall into two categories: 1) Calcium alginate method and, 2) Alginic acid method. The chemistry of the processes used to make sodium alginate from brown seaweeds is relatively simple. The difficulties of the processes arise from the physical separations which are required, such as the need to filter slimy residues from viscous solutions or to separate gelatinous precipitates which hold large amounts of liquid within the structure and which resist filtration and centrifugation.

Commercial Uses of Sodium Alginate and Applications in Food

There are numerous applications for alginates in foods. It is used as a stabilizer for ice cream, yogurt, cream, and cheese. In ice cream making, the addition of sodium alginate reduces the formation of ice crystals during freezing, giving a smooth outcome.

It is utilized in fruit-pulp texturization, diffusion-set gels, protein extrusion, storage-life extension of potatoes, entrapment of enzymes, inclusion in fish patties, and use in structured beef products.

It acts as a thickener and emulsifier for salad, pudding, jam, tomato juice, and canned products. It is a hydration agent for noodles, bread, cool and frozen products. In the presence of calcium and acid mediums, it forms resilient gels. It is a cold gelling agent that needs no heat to gel. It is most commonly used with calcium lactate or calcium chloride in the spherification process.

In the commercial food industry, sodium alginate is used as thickener in sauces, and syrups. By thickening pie fillings with sodium alginate, softening of the pastry by liquid from the filling is reduced. Water-in-oil emulsions such as mayonnaise and salad dressings thickened with sodium alginate are less likely to separate into their original oil and water. The use of alginates is expected to increase in the near and distant future.

Used in Making Alginate Spheres

The most common application of sodium alginate in molecular gastronomy or culinary physics is to create spheres using the spherification technique made famous by Chef Ferran Adria at el Bulli restaurant in 2003. Spherification is the culinary process of shaping a liquid into spheres which visually and texturally resemble caviar. The technique was originally discovered by Unilever in the 1950s. The resulting spheres have a thin membrane of gel and are filled with liquid.

Culinary Physics or molecular gastronomy is a method of applying scientific principles to cooking or baking. When mixed in the right proportions before the addition to the calcium solution, the substance is found to be in liquid form. When expunged through a syringe, drop by drop into the calcium solution, the liquid forms a gelatin coating and resembles individual caviar.

Alginates are Used in Fish and Meat Preservation

Block-freezing fatty fish, such as herring or mackerel, in alginate jelly might prevent oxidative rancidity. An alginate film that excludes air is formed around each fish piece, making rancidity almost impossible. Storage with the coating also assists in decreasing the off-flavors and unpleasant odors correlated with fish. This procedure has been extended to other fish, as well as shellfish. Herring can also be dusted with alginates to assist in its canning. The film-forming sodium alginate and tamarind kernel powder, guar gum, and agar–agar may also be beneficial for shelf-life extension and curing preservation of salted and dried mackerel.

Highly water-soluble fish-meat protein can be stably manufactured by conjugation with alginate oligosaccharides. Frozen mature salmon meat can be used as the raw material for a highly stable meat–alginate oligosaccharide conjugate. In addition, inhibition of protein denaturation in the processing line is vital to manufacturing a meat–alginate oligosaccharide conjugate with high water solubility. The water-soluble fish meat might be used to develop new food items such as protein rich beverages, nutritional emulsifiers, and a source of protein for people who cannot chew foods.

Poultry parts have been coated with calcium-alginate films. Whole beef cuts were also coated with calcium-alginate films prior to freezing, by dipping in a 10–15% sodium-alginate solution, then 3.5–5% calcium chloride, followed by 10–20% glycerol (as a plasticizer). An edible sodium alginate–cornstarch meat coating was used to coat beef steaks, pork chops, and skinned chicken drumsticks to improve color, appearance, odor, texture, and juiciness. Raw poultry items can serve as a source of human pathogens that may cross-contaminate additional foods.

Despite efforts to control them during rearing, shipping, and processing, elimination of poultry contamination by these organisms presents a significant challenge. To improve effectiveness without changing processing speeds at the plant, edible gels based on pea starch and calcium alginate containing antimicrobials can be applied to chicken surfaces. Theoretically, the antimicrobial agents will slowly diffuse from the gels or coating material into skin irregularities, and application after de-feathering will provide increased contact time with target microorganisms and yield improved effectiveness.

An edible alginate coating was reported that improves the quality of frozen pork. The coating decreased meat loss during thawing, and was very useful in maintaining the functional properties of the frozen meat. In addition, no significant differences in the structures of the control and coated pork were observed.

Used in Dairy based Drinks and Yogurt

Sodium alginate is often found in chocolate milk, eggnog, drinkable yogurt and fruit-flavored yogurts to make a smoother mixture, as a clarifying agent and as a stabilizer.

Sodium Alginate in your Favorite Cheese

A minute addition of sodium alginate to soft cheese spreads can prevent separation of water and oil, diminish surface hardening, and produce a better textured processed cheese. Inclusion of ~0.15% sodium alginate is adequate to thicken whipped cream, that is, to act as a stabilizer upon whipping. The manufacture of synthetic creams can be facilitated by the addition of alginate to achieve a quicker whip, greater tolerance to over whipping, stabilized overrun, and prevention of syneresis.

Alginates are also Used in Sausage Casings

Alginates were also used as part of restructured meat products and to produce synthetic sausage casings. In 1955, the Visking Corporation pioneered alginate casings under the name of Tasti- Jax. The casings were not successful in the United States but were extensively used in Germany.

A characteristic casing was made of 6% sodium alginate extruded into an acidic solution of 10%

calcium chloride, which was then plasticized by washing with glycerol and calcium lactate. On the one hand the casings were elastic, hygienic, and easier to handle than natural casings. On the other, likely due to interaction with sodium salts, the casings became swollen, weaker, and less attractive, and did not succeed in contracting during cooking. Attempts to eliminate these shortcomings were suggested. Extruded edible casings manufactured from blends of sodium alginate with pectin or gelatin for use with breakfast pork sausage were assessed. Sensory analysis demonstrated a preference for the pectin casings over the gelatin ones.

The quality of low-salt sausages enriched with inulin was found to vary when some of the pork's back fat was replaced with olive oil. The addition of an olive oil–alginate emulsion and inulin resulted in a low-salt, reduced-fat product that was richer in monounsaturated fatty acids, while retaining sensory notes similar to those of the traditional sausage used as a control; this product achieved a good acceptability rating.

Sodium Alginate is Used in Making Fruit-like Products and Coatings on Fresh-cut Fruit Pieces

Fruit and other texturized products can be prepared from alginates. Simulated fruits can be effectively produced by means of a two-step mixing internal-setting procedure. The first phase consists of mixing together alginate, anhydrous dicalcium phosphate, disodium hydrogen ortho-phosphate, glucose, sucrose, and water. In parallel, fruit purée, sucrose, glucose, citric acid, and sodium citrate dihydrate are also blended. At neutral pH, the anhydrous dicalcium phosphate is insoluble. Upon pumping the two mixtures, first under high-shear mixing, then under no-shear conditions, the lowered pH causes calcium ion release from the anhydrous dicalcium phosphate leading to a gelling reaction. This process can also be executed in a continuous manner.

There are many recipes and patents that deal with the formation of texturized fruits. One example is the Unilever patent (BP 1,484,562), which describes a co-extrusion system for preparing reformed black currants. In general, fruit purées offer new outlets for visually imperfect or excessively small fruit. To restructure the purée, a texturizing agent such as alginate is required to control the functional properties of the product. Novel gel systems were formed from alginates and peach purée, without supplementary calcium or sugar. These gels were shown to be the product of interactions between pectin and alginate. Novel alginate–peach gels involved minimal preparation and could be scaled up in a very straightforward manner and employed by the food industry for a variety of end products: final fruit products that contain up to 99% fruit and are highly nutritious, and have commercial potential as ready-to-eat snack foods or as constituents of baked, frozen, and/or canned foods.

Coatings based on alginate or gellan gum were used on fresh-cut papaya pieces to preserve the quality of the minimally processed fruit. Formulations including glycerol and ascorbic acid revealed somewhat improved water-barrier properties as compared to the uncoated samples. Inclusion of sunflower oil in the formulations increased the water-vapor resistance of the coated samples. The coatings also improved the firmness of the fresh-cut product, and the inclusion of ascorbic acid as an antioxidant in the coatings helped maintain the cut fruit's nutritional quality throughout storage.

Sodium Alginate in Ice Cream and Popsicles

Sodium alginate is used in ice cream to assure a creamy texture and prevent ice crystals from forming. In fruit flavored popsicles, sodium alginate helps distribute fruit uniformly during the freezing process and stops them from dripping while eating.

Alginates are Used in Bakery Toppings, Fillings, Beverages and Salad Dressings

Alginates have been used to produce non sticky icings, which do not crack. Alginates also successfully maintain the stability of foam and aerated structures and improve the texture of whipped sugar toppings. Alginates can minimize syneresis in baking jellies and enhance their stability at elevated baking temperatures. Pie fillers are stabilized with alginates. Moreover, better clarity of frozen and canned cherry and peach pastry fillings has been achieved by combining PGA with a modified waxy cornstarch.

Fruit-pulp precipitation in drinks can be delayed by inclusion of PGA or sodium alginate within the formulation. PGA is also an efficient stabilizer in fermented fruit milk drinks. An alginate–phosphate blend is useful as a stabilizer in chocolate-milk drinks. Another combination of pectin, polyol alginate and a fatty acid ester stabilizes acidic drinks composed of an aqueous blend of vegetable extract, juice, and milk products. PGA might help in stabilizing beer foam, and sodium alginate can be utilized to clarify wine and eliminate tannins, nitrogenous substances, and coloring material.

Dressings are in essence vegetable-oil emulsions, which also include salt, sugar, spices, and flavorings. Sodium alginate and PGA slow separation of the oil and water phases, thus stabilizing the dressings or sauces. Gum tragacanth can also be used for this purpose. Inclusion of PGA with cornstarch in salad dressings improves their stability, producing a final soft product in which the oil does not separate out upon standing. PGA and xanthan gum stabilize oil-in water emulsions in the presence of salt as a dissolved electrolyte. Low-salt concentrations are particularly effective when PGA–xanthan gum combinations are used in the presence of polysorbate-60.

Alginates are Used in Manufacture of Milk Puddings and Ice-Cream Stabilizers

The manufacture of milk pudding based on a specifically treated mix of water soluble, alkali-metal alginate, a mild alkali, and a minute amount of calcium salt has been reported. A simple mix was used to form a good gel in water, which could be admixed with powdered skim milk to structure a milk pudding when dissolved in cold water. Ice-crystal growth in ice creams can be slowed by alginate to achieve a smooth texture. An identical addition can also contribute to delaying product melting and controlling overrun. Minute quantities (i.e., 0.1 to 0.5%) of sodium alginate are most frequently used as ice-cream stabilizers. Given that the stabilizers comprise water-holding properties and dispersive capabilities, they add to high-quality body properties and texture protection. Sodium alginate may react with the calcium in the ice-cream mix, thereby decreasing the concentration of calcium ions in the water. This is also useful for getting rid of the clumping of fatty globules in the manufactured goods.

In Making Sauces, Gravies and Dressings

Sodium alginate is used to stabilize liquid packaged salad dressings, sauces and gravies to extend their shelf life and prevent separation.

Used in Water Dessert Gels

Eye-catching edible gels can be manufactured with alginates. Calcium is frequently the cation of choice for cross-linking; nevertheless, divalent or trivalent metal ions can also be employed. It is suggested that the calcium ion concentration and pH be properly selected to gain more control over reaction rates and the texture of the formed product. Gel formation that proceeds too rapidly creates a grainy gel, whereas extremely slow gelation results in very soft gels. Controlled release of calcium ions can be achieved by using a suitable salt to manage its solubility and by using sequestering or retarding agents. Such systems have been the subject of many patents and have been engaged for the manufacture of dessert gels and candied jellies, fruit jams and jellies, and jellied salads and broths.

Other Food Applications of Alginates

Alginate gels could be useful in numerous other applications. Yeast cells have been immobilized in alginate beads for alcoholic drinks and ethanol production and in the second fermentation of champagne. Alginate thermo stability could be beneficial for microwaving alginate-based products. In addition, alginates enable the fabrication of novel fried foodstuffs that have better taste and texture and are more easily adapted for mass production. PGA is beneficial in the manufacture of dietetic dressings. Alginates have also been used in macaroni, spaghetti, and dough manufactured goods to allow a higher rate of extrusion.

Gellan Gum

Gellan gum is a food additive typically used to bind, stabilize, or texturize processed foods. It's similar to other gelling agents, including guar gum, carrageenan, agar agar, and xanthan gum. It grows naturally on water lilies but can be artificially produced by fermenting sugar with a specific strain of bacteria.

It's a popular replacement for other gelling agents because it's effective in very small amounts and produces a clear gel that isn't sensitive to heat. Gellan gum also works as a plant-based alternative to gelatin, which is derived from animal skin, cartilage, or bone.

How is Gellan Gum Used?

Gellan gum has a variety of uses:

- As a gelling agent, it lends a creamy texture to desserts, gives fillings for baked goods a jelly-like consistency, and reduces the likelihood that certain delicacies — such as creme brûlée or flaming sorbet — will melt when subjected to heat.
- Gellan gum is also commonly added to fortified juices and plant milks to help stabilize supplemental nutrients like calcium, keeping them mixed into the beverage rather than pooled at the bottom of the container.
- This additive likewise has medical and pharmaceutical applications for tissue regeneration, allergy relief, dental care, bone repair, and drug manufacturing.

Foods that Contain Gellan Gum

You can find gellan gum in a variety of foods, including:

- Beverages: Fortified plant-based milks and juices, chocolate milk, and some alcoholic drinks.
- Confectioneries: Candy, marshmallows, fillings for baked goods, and chewing gum.
- Dairy: Fermented milk, cream, yogurt, processed cheese, and some unripened cheeses.
- Fruit and vegetable products: Fruit purées, marmalades, jams, jellies, and some dried fruit and vegetables.
- Packaged foods: Breakfast cereals, as well as some noodles, potato gnocchi, bread, rolls, and gluten-free or low-protein pastas.
- Sauces and spreads: Salad dressings, ketchup, mustard, gravies, custards, and some sandwich spreads.
- Other foods: Some processed meats, fish roe, soups, broths, condiments, powdered sugar, and syrups.

Gellan gum is particularly popular in vegan packaged foods because it's a plant-based alternative to gelatin.

Pectins

Commercial pectins are prepared mostly from some by-products of the food industry, such as apple pulp, citrus peels, and sugarbeet pulp.

Pectin is the major constituent of all plants and makes up approximately two-third of the dry mass of plant primary cell walls. It provides structural integrity, strength, and flexibility to the cell wall and acts as barrier to the external environment. Pectin is also a natural component of all omnivorous diet and is an important source of dietary fiber. Due to the resistant in digestive system and lack of pectin digestive enzymes, human beings are not able to digest pectin directly but microorganism present in large intestine can easily assimilate the pectin and convert it into soluble fibers. These oligosaccharides promote beneficial microbiota in gut and also help in lipid and fat metabolism, glycemic regulation, etc.. Being complex and highly diverse in structure, role of pectin is not only limited to the biological and physiological functions, but it has tremendous potential and contributes substantially in other applications ranging from food processing to pharmaceuticals. Pectin is a water-soluble fiber and used in various food as emulsifier, stabilizer, gelling, and thickening agent.

Commercial pectins are extracted from citrus and apple fruit. On the basis of dry mass, apple pomace contains 10–15% pectin, whereas citrus peel possesses 20–30% pectin. However, pectin has also been extracted in higher amount from several other fruits and their by-products, such as sunflower head, mango peal, soybean hull, passion fruit peel, sugar beet pulp, *Akebia trifoliata* peel, peach pomace, banana peel, chickpea husk, and many more. Table summarizes the different types of pectin extracted from various horticultural crops. But detection and extraction of pectin in higher concentration is not sufficient to qualify that fruit as a source of commercial pectin because of the structural variation and modification in side-chain sugars, and also that pectin from different sources has different gelling properties.

S. No	Source	Parts used	Extraction meth od used	Pectin yield (%)	Type of pectin (HMP/LMP)
1	Passion fruit	Peel	APP	14.8%	HMP
2	Banana	Peel	APP	5–21%	HMP (DE, 50–80%)
3	Chick pea	Husk	Acid extraction, APP, and freeze dried	8%	LMP (DE, 10%)
4	Krueo Ma Noy	Leaves	APP, DPP	21–28%	LMP (DE, 34–42%)
5	Yellow Passion	Fruit rind	APP, DPP, MPP	3–16%	HMP (DE, 54–59%)
6	Durian	Rind	APP	2–10.25%	HMP (DE, 50–64%)
7	Mulberry	Mulberry bark with epidermis (MBE) and without epidermis (MB)	Extracted using 60–100% isopropanol	11.88%	HMP (MB–DE, 71.13%); LMP (MBE–DE, 24.27%)
8	Yuzu, citrus family	Pomace	Extracted with APP and enzyme (Viscozyme L with 1.2×10^{-4} fungal β-glucanase	DPP, APP (7.3–8%)	LMP (APP–DE, 41%; DPP–DE, 46.3%)
9	Cacao pods	Husk	Extracted with 1 N HNO_3 at different pH and precipitated by ethanol and acetone	3.7–8.6%	LMP (DE 36.7% @ pH 1, DE 44.3% @ pH 3); HMP (DE 52.4% @ pH 2)
10	Cashew apple	Pomace	AOP at different pH (1.0, 1.5, and 2.0)	10.7–25.3%	LMP (DE, 28–46%)
11	*Cyclea barbata* Miers (CBM)	Leaves	Extracted with acid and alkali, precipitated the pectin by ethanol	4–8%	HMP (acid treated: 65–75% DE) LMP (Alkali treated: 36% DE)
12	Dragon fruit	Peel	Extracted using HCl, precipitated and purified with 70 and 99.6% isopropanol.	18.59%	LMP (DE, 46.95%)
13	Jackfruit	Peel	Ultrasonic-microwave-assisted extracted (UMAE) pectin	21.5%	HMP (DE, 62.5%)
14	Potato	Pulp	Extracted with different acids and precipitated by ethanol.	4.08–14.34%	LMP (DE, 21.51–37.45%)

High methoxyl pectins (HMP) and low methoxyl pectins (LMP) from various horticultural crops. APP, alcohol-precipitated pectin; MPP, metal ion-precipitated pectin; DPP, dialyzed precipitated pectin.

Pectin Structure

Pectin is a highly complex plant cell wall polysaccharide that plays a significant role in plant growth and development. It is predominantly present in fruits and vegetables and constitutes approximately 35–40% of the primary cell wall in all the dicot plants. The composition and structure of pectin is influenced by the developmental stages of plants. Structural analysis of pectin revealed that it is a polymer comprised of chain-like configuration of approximately 100–1000 saccharide units; therefore, it does not possess a defined structure. In general, pectin is illustrated as a heteropolysaccharide of three components namely, homogalacturonan (HG), rhamnogalacturonan-I (RGI), and rhamnogalacturonan-II (RGII). The Backbone structure may branch with other neutral sugar chains such as arabinan, xylogalacturonan (XGA), arabinogalactan I (AG-I), and arabinogalactan II (AG-II).

Homogalacturonan (HG) is a polymer of galacturonic acid (GalA), in which Gal A residues are linked together by α-1-4 glycosidic bond and the number of GalA residues in HG may vary from 72 to 100% depending on the source of pectin. For instance, the HG backbone of cashew apple pectin, C. maxima pectin, sunflower pectin, citrus pectin, comprises of 69.9–85%, 71–75%, 77–85%, 80–95%, GalA residues, respectively. Amaranth pectin contains more than 80% GalA residues in HG backbone structure. Furthermore, it was also observed that HG may be methoxy-esterified at C-6 and/or O-acetylated at the O-2 and/or O-3. Some exception has also been reported in the detailed structural analysis of HG region of pectin such as C-3 substitution of the galacturonic acids of HG with xylose in pea, apple, carrot, duck weed, etc., and C-2 or C-3 with apiose in duck weeds (Lemna minor). HG is susceptible to both mechanical and enzymatic deesterification and degradation.

Rhamnogalacturonan I represents approximately 20–35% of the pectin polysaccharides. It is the highly branched and heterogeneous polysaccharide which is characterized as repeating units of α-(1 → 2)-linked rhamnose and α-(1 → 4)-linked GalA residues. It can be O-acetylated at O-2 and/or O-3 positions of GalA residues. Pectin from citrus peels, mung bean, kidney bean, apple fruit, and flax hypocotyls has been reported 100% methyl esterified in the RGI region. The composition of RGI varies in pectin extracted from different sources. In sugar beet pectin, 80 repeating units of [→2] –α-L-Rha-(1–4)- α-D-GalA-(1→) comprised the backbone of rhamnogalacturonan I (RG-I), whereas citrus pectin contains only 15–40 repeating units. The polymeric side chains of galactans and arabinans are substituted at the O-4 position of RG-I backbone. Arabinogalactan I (AG-I) and arabinogalactan II (AG-II) are also reported to be present as polymeric side chains. The side chains are often referred to as "hairs" and believed to play an important role in pectin functionality. The loss of side chains may increase the solubility of the pectin. PGI is prone to enzymatic de-polymerization. However, protease and acid-catalyzed cleavage of RGI has also been reported.

The highly conserved polysaccharide of pectin is rhamnogalacturonan II which constitutes about 10% of the pectin polymer. This polysaccharide is made up of (1 → 4)-linked-α-D-GalA units containing 12 monosaccharide such as apiose, acetic acid, 3-deoxy-manno-2-octulosonic acid (KDO), and 3-deoxy-lyxo-2-heptulosaric acid (DHA) as side chains. GalA present in backbone of rhamnogalacturonan II (RG-II) may be methyl esterified at the C-6 position. The percentage of esterified GalA and acetylated groups in HG chain is termed as the DE and DAc, respectively. It is proposed that in the early developmental stages of plants, highly esterified pectin is formed that undergoes some deesterification in the cell wall or middle lamella. In general, tissue pectin ranges from 60 to 90% DE. Both the DE and the DAc of pectin may vary depending on the method of extraction and plant origin. The functional properties of the pectin are determined by the amount and the distribution of esterified GalA residues in the linear backbone. Presence and distribution of esterified and nonmethylated GalA in pectin define the charge on pectin molecules. Based on their degree of esterification (DE), pectins are classified as high methoxy pectins (HMP) or low methoxy pectins (LMP). DE values of HM pectin range from 60 to 75%, whereas pectin with 20–40% of DE is referred as LM pectin. It was also observed that solubility, viscosity, and gelation properties of pectin are correlated and highly dependent on structural features. Pectin and monovalent salts of pectins are generally soluble in water but di- and trivalent ions are insoluble. The solubility of pectin in water increases with decrease in polymer size and increase in methoxy contents. Pectin powder gets hydrated very fast in water and forms clumps. The solubility of these clumps is very slow. As the pectin molecules come in contact with water, deesterification and depolymerization of pectins start spontaneously. The rate of decomposition of pectin depends on pH and temperature of the solution. As the pH of the solution decreased, with elevated temperature, ionization of carboxylate groups also reduced, which suppresses the hydration and repulsion between the polysaccharide molecules and results in the association of molecules in the form of gels. During thermal processing, solubilization of pectin is affected by β-elimination which depolymerized the pectin molecule and reduced its chain length. Small polymers have poor affinity with cell wall framework and solubilize easily. However, preheating, as well as reduced moisture contents in thermal processing, adversely affects the solubility of pectin in water.

Pectin as Food Emulsifier

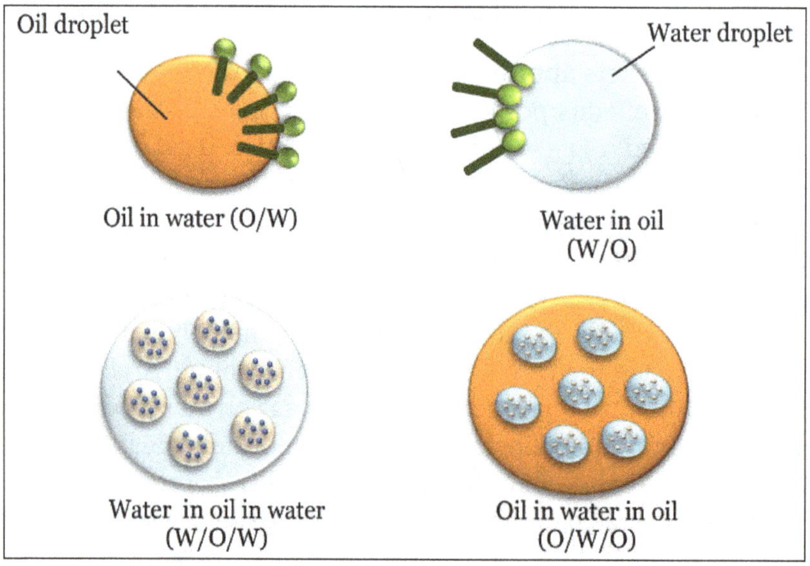

Types of emulsions.

Food additives that are used in food processing to blend two immiscible liquids to produce a desirable product are known as food emulsifier or emulgent. These additives act as surface-active agents on the border of immiscible layers and reduce oil crystallization and prevent water separation. Emulsifiers are used in large number of food products such as ice creams, low-fat spreads, yoghurts, margarine, salad dressings, salty spreads, bakery products, and many other creamy sauces, to keep them in stable emulsion. Emulsifiers increase the whip-ability of batters, enhance mouthfeel of the products, and improve texture and shape of the dough. Moreover, emulsions also help to encapsulate the bioactives. Based on the disperse phase, there are two types of emulsion: oil in water (O/W) and water in oil (W/O). Milk, mayonnaise, dressings, and various beverages are some examples of O/W emulsion, whereas butter and margarine are the typical examples of W/O emulsion. Progress in hydrocolloid chemistry has resulted in the development of multitype emulsion such as O/W/O and O/W/O type emulsion. These emulsions are very important for fat reduction or encapsulation of bioactives and are used in preparation and stabilization of various low-fat creams, seasoning, and flavoring of sauces.

Commonly used emulsifiers in food processing are (i) small-molecular surfactant such as lectithins, derivatives of mono- and diglycerides prepared by mixing edible oils with glycerin or ethylene oxide, fatty acid derivatives such as glycol esters, sorbitan esters, polysorbates and (ii) macromolecular emulsifiers that include proteins and plant-based polymers such as soy polysaccharide, guar gum, modified starch, pectin, etc. As far as the properties of food emulsifier are concern, a good emulsifier should be low in molecular weight, capable to reduce the surface tension rapidly at interface, and should be soluble in continuous phase. Research on food additives revealed the adverse effect of synthetic food additives on human being. Chassaing et al. found that polysorbate 80(P80) or carboxy methyl cellulose (CMC) had adverse effects on gut microbiota and their continuous use triggered the weight gain and metabolic syndrome after 12 weeks of administration in mouse. Therefore, safety issues with the synthetic food additives and consumer's demand for all natural food ingredients have necessitated the use of plant-based emulsifiers and stabilizers in food.

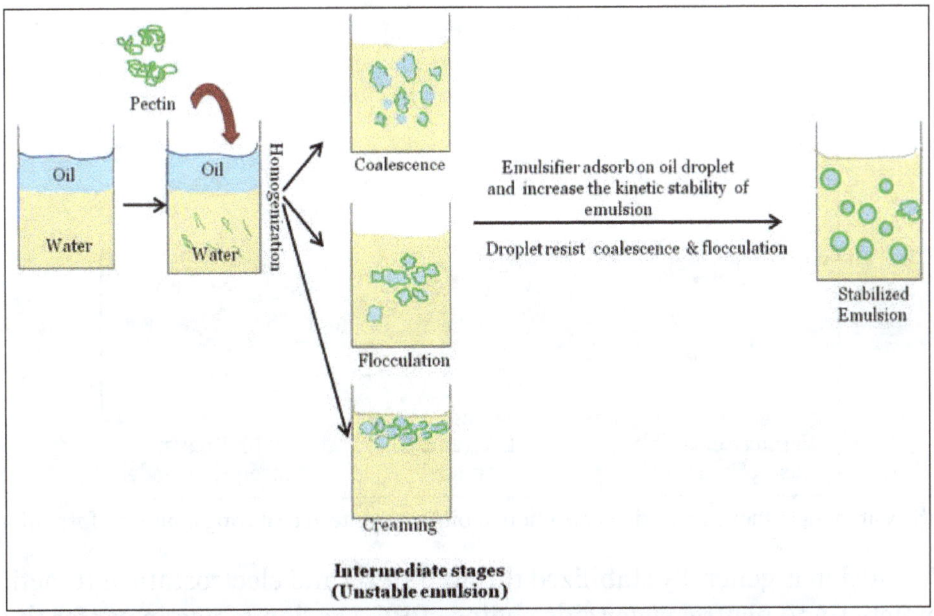

Emulsion formation and stabilization using polymer as emulgent.

Pectin is a natural hydrocolloid which exhibits wide spectrum of functional properties. Because of the gelling ability of pectin, it is used as viscosity enhancer. During emulsification process, pectin molecules adsorb at the fine oil droplets from at O/W interface and protect the droplet from coalescing with adjacent drops (short-term stability). The quality of emulsifier is defined by its ability to provide long-term stability against flocculation and coalescence. Figure depicts the stages in long-term emulsion formation using pectin as emulgent. When the viscosity of the continuous phase is increased, the movements of oil droplets become restricted which improves the shelf life of emulsion. In the past decade, some pectin has also been reported to exhibit surface active behavior in oil-water interface and thereby stabilizing the fine oil droplets in emulsion. These functions of pectin are determined by its source, structural modification during processing, distribution of functional groups in pectin backbone, and also by various extrinsic factors such as pH, temperature, ionic strength, cosolute concentration, etc. The emulsification or surface active properties of pectin, i.e., formation of fine oil droplets, are mainly contributed due to the high hydrophobicity of protein residue present in pectin and also by hydrophobic nature of acetyl, methyl, and feruloyl esters, whereas emulsion-stabilizing ability is attributed to the carbohydrate moieties and their conformational features.

Mechanism of Emulsion Formation and Stabilization

The mechanism of emulsion formation is shown in Figure. Different models explain the emulsion formation as covalently bound protein moieties in pectin are adsorbed onto the oil-water interface, form anchor points at the interface, and reduce the interfacial tension while the charged carbohydrate units extend into the aqueous phase and stabilize by steric and viscosity effects in the aqueous phase. Now, it is a well-established fact that pectin from different source shows variability in structure and protein contents. Leroux et al. identified many anchor points in sugar beet pectin (SBP) molecules, and proposed a loop-and-tail model.

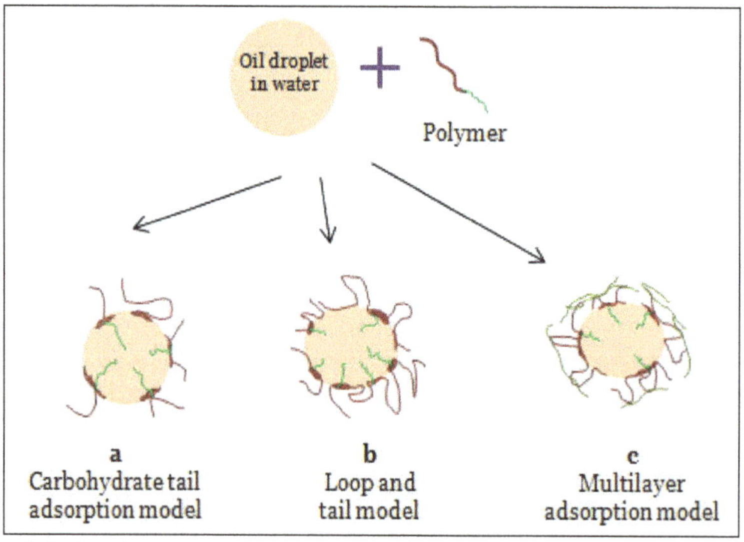

Different models showing pectin adsorption at oil/water interface during emulsion formation.

Pectin O/W emulsion is generally stabilized through steric and electrostatic interaction. The carbohydrate moieties and neutral sugar side chains of RG I region of pectin confer the stability to the pectin emulsions through steric properties of the adsorbed polymers, when pectin is used as monoemulsifiers. In addition, pectin reversible association with galactan/arabinogalactan prior to emulsification also improves the emulsion stability. Electrostatic stabilization of emulsion is ascribed to sugar moieties and structural features of the HG units of pectin. If the pH of dispersion medium is above 3.5, nonmethylated carboxylic group of HG region gets ionized and confers charge on the pectin surface. Interaction of an ionic surfactant with oil droplets results in electrostatic stabilization. Pectin viscosity also plays an important role in controlling the emulsion stability. HG region-rich pectin shows higher intrinsic viscosity ([η]); therefore, HG and RG ratio of pectin and molecular interactions that improve the intrinsic viscosity ([η]) of pectin solution also contributes in shelf life of emulsion. It has also observed that structural features of pectin such as pectin protein content, molecular mass, and presence of ferulic acid, and acetyl group in carbohydrate moieties of pectin also affect pectin's emulsifying and emulsion stabilization properties. Ferulic acid-rich pectin did not show significant difference in emulsifying ability of pectin when compared with pectin poor in ferulic acid. Digestion of sugar beet pectin(SBP) with acidic proteases resulted in formation of larger size of oil droplet, lower creaming stability, and loss of emulsifying activity of SBP which confirms that protein contents of SBP play an important role in emulsifying ability of the polymer. It was also found that protein-rich fractions of SBP did not necessarily displayed better emulsifying ability; therefore, it was concluded that both protein with carbohydrate moiety together help in controlling emulsifying ability of SBP. Both the carbohydrate and protein moieties function together as unit and affect the hydrophilic-hydrophobic equilibrium of the SBP molecule. Therefore, when SBP is digested with proteases or other enzyme, a single moiety may function differently. Furthermore, it was also proposed that protein folding may also mask the hydrophobic effect of protein and thus affect the overall properties of the polymers.

Molecular weight of pectin has also been reported to affect the emulsifying capacity of pectin. Pectin with low molecular weight was more efficient in stabilizing small emulsion droplets than high-molecular weight pectin. However, very small size of citrus pectin had negative effect on emulsion-stabilizing ability of pectin. It could be due to the poor steric stabilization of depolymerized polymer.

Pectin Containing Emulsion-based Food

Emulsion-based food products can be defined as a network of pectin-protein molecules entrapping the oil droplet in between. Nowadays, a large number of pectin- and polysaccharide-based emulsified low-fat dairy products, meat products, spreads or desserts, bakery products, sauces, etc., are available in market. Low-fat and low-cholesterol mayonnaise, low-fat cottage cheese, low-fat drinking yogurt, and flavored oil-containing acidified milk drinks are the few examples of pectin-based emulsified products. These products are prepared by replacing full-fat milk from skimmed milk, emulsified oil, and whey proteins. A low-fat cheese was prepared using skimmed milk and water-in-oil-in-water ($W_1/O/W_2$) emulsified canola oil. Different emulsifiers such as amidated low-methoxyl pectins (LMP), gum arabic (GA), carboxymethylcellulose (CMC), and combinations of GA-CMC or GA-LMP were used to stabilize the emulsion. Textural characteristics and sensory evaluation of low-fat cheese show that polymers used to stabilize the emulsion affected both microcrystalline structure and organoleptic properties. The cheese prepared using GA and LMP was almost similar in textural characteristics to the full-fat milk cheese. In another study, Liu et al. compared the textural and structural features and sensory quality of full-fat and low-fat cheese analogs prepared with or without the incorporation of pectin. Microstructure analysis using scanning electron microscopy revealed that full-fat cheese was denser and contained higher concentration of fat globules than low-fat cheese made with or without pectin. Comparison within the low-fat cheese analogs showed clear difference in their hardness, gumminess, chewiness, and adhesiveness. Addition of pectin had positive effect on textural and sensory attribute and scored better in mouthfeel also.

Low-fat (Lf) mayonnaise was prepared by partial replacement of egg yolk and incorporation of pectin as emulsifier. Pectin weak gel, pectin microencapsulation, and whey protein isolate were used in preparation of low-fat (Lf) mayonnaise. Physicochemical and sensory properties of Lf mayonnaise were compared with full-fat (Ff) mayonnaise; Lf mayonnaise had low energy and more water contents than Ff. Textural features and rheological properties of the Lf and Ff mayonnaise were similar and both displayed thixotropic shear thinning behavior and categorized as weak gels. Moreover, Lf mayonnaise prepared using pectin had better acceptability than whey protein incorporation. Emulsified oil is used as an effective delivery system of active compound in functional foods, and also serves as milk fat replacer in fat-free dairy products. To improve the nutritional value of food, low-fat dairy products are produced, whereas saturated milk fat is generally replaced with emulsified-unsaturated vegetable oils.

In recent year, pectin in combination with inulin has been reported to prepare low-fat meat batter. Méndez-Zamora et al. studied the effect of substitution of animal fat with different formulations of pectin and inulin on chemical composition, textural, and sensory properties of frankfurter sausages. Fracturability, gumminess, and chewiness of the low-fat sauces were slightly lower than those of the control. However, addition of 15% inulin improves the sensory properties. In a similar work, replacement of pork back fat with 15% pectin and 15% inulin was found effective in maintaining the physicochemical properties and emulsion stability of the low-fat meat batter.

Pectin as Gelling Agent

The use of pectin in food products as a gelling agent is a long tradition. Later on, it was discovered that pectin forms different types of viscoelastic solution under suitable conditions. This property of pectin is commercially exploited in preparation of jams, jellies, and marmalades. Rheological behaviors of pectin depend on pectin source, its degree of methylation, distribution of nonmethylated GalA unit on pectin backbone, and degree of acetylation, and also on various extrinsic factors such as temperature, pH, concentration, and presence of divalent ions. At a constant pH, the setting time of pectin increases with decreasing DM and degree of blockiness (DB) in the absence of bivalent ions. Therefore, on the basis of gelling process, pectin is classified as rapid, medium, and slow set pectin.

Gelling process of pectin and its stabilization follows different mechanisms for different types of pectin. HMP form gels in a narrow pH range (2.0–3.5) in the presence of sucrose at a concentration higher than 55% w/v in medium. During the gelatin process of HMP, junction zones are formed due to the cross-linking of two or more pectin molecules. These junctions are stabilized by weak molecular interaction such as hydrogen and hydrophobic bonds between polar and nonpolar methyl-esterified groups and require high sugar concentration and low pH. These gels are thermally reversible. LMP can form gel over a wide pH range (2.0–6.0) independent of sucrose, but requires divalent ion, such as calcium. LMP follow the eggbox model for its gelation, where positively charged calcium ions (Ca^{2+}) are entrapped in between the negatively charged carboxylic group of pectin. The zigzag network of Ca^{2+} ion and GalA molecules looks like eggbox, and therefore, model is named as eggbox model. These gels are stabilized by electrostatic bonds. In the presence of Ca^{2+}, calcium bridges are formed with pectin molecules that make the solution more viscous. At the higher pH, the ionic strength of the solution is increased and thus more Ca^{2+} is needed for gelation. In case of highly acetylated pectin such as sugar beet, acetyl groups cause steric hindrances and interfere with the Ca^{2+} ion and GalA bond formation, thus preventing gel formation. Enzymatic modification in pectin structure, i.e., removal of acetyl groups using α-arabinofuranosidase (α-Afases) and acetyl esterase enzymes, can improve the gelling property of acetylated pectin.

HMP are generally used in preparation of standard jams where sugar contents are above 55%, high-quality, tender confectionary jellies, fruit pastes, etc. LMP do not require sugar for its gelatin and therefore preferred choice for the production of low-calorie food products such as milk desserts, jams, jellies, and preserves. LM pectins are more stable in low pH and high temperature conditions as compare to HM pectins and can be stored for more than a year.

Pectin in Food Packaging

Food packaging is one of the fastest growing segments of food industry. Traditionally, packaging system was limited to the containers and packaging material to transport the food items from manufacturer to the retail market and then to the consumers. Such type of packaging was unable to contribute in the extension of the shelf life and maintenance of the quality of the products. Due to the globalization of food market and increasing demand of shelf-stable processed food that retains the natural properties of food, the need of functional/active packaging material is increasing. To meet the industrial demand, a number of polymers are being synthesized and used in food packaging because of their flexibility, versatility, and cost effectiveness. Although, synthetic materials are able to fulfill all the industrial needs and keep food fresh and safe by protecting them from abiotic factors such as moisture, heat, oxygen, unpleasant odor, and biotic components such as micro- and macroorganisms. But, disposal of nonbiodegradable packaging material is a serious problem which poses a threat to the environment.

A variety of natural polysaccharides, such as pectin, chitosan derivatives, alginate, cellulose, seaweed extract, and starch are usually used in the preparation of edible films and coatings. Pectin is one of the most significant renewable natural polymers which are the main component of all the biomass and ubiquitous in nature. Being flexible in nature, pectin and its derivatives are used in many biodegradable packaging materials that serve as moisture, oil, and aroma barrier, reduce respiration rate and oxidation of food. Pectin along with food grade emulsifiers is also used in the preparation of edible films. These films are used in fresh and minimally processed, fruits and vegetables, foods and food products as pectin is the main component of the omnivorous diet and can be metabolized. Edible coating protects the nutritional properties of the food and also saves highly perishable food from the enzymatic browning, off-flavor development, aroma loss, retards lipid migration, and reduces pathogen attack during storage.

At low pH, LM pectins are cross-linked with calcium cations and form hard gels. These gels have highly stable structure and act as water barriers. Because of these properties, LM pectin films are used as edible coatings. Extension of shelf life of avocado fruits was also reported to over a month at 10 °C by using edible pectin films. It was found that when avocados were coated with edible pectin films and stored at 10 °C, rate of oxygen absorption and rate of respiration decreased which results in delaying of texture and color change of fruits. Oms-Oliu et al. used calcium chloride and sunflower oil cross-linked with LM pectin films onto fresh-cut melon to see the effect on extension of shelf life of cut fruits. It was observed that edible pectin films maintained the initial firmness, decrease the wounding stress of fresh-cut fruits, and prevent the dehydration during storage up to 15 days at 4 °C but could not reduce the microbial growth onto the fresh melon. It has been observed that to reduce the respiration rate and to prevent the off-flavor development, different pectin and emulsifier formations are required for different fruits. Edible coating film formulation consisted on pectin, sorbitol, and bee wax was successfully used by Moalemiyan et al. to keep the fresh-cut mangoes in original state for over 2 weeks. Whereas in a similar study, pectin coating containing sucrose and calcium lactate was able to prevent the fruits' respiration rate and maintain sensory properties in fresh melon fruits for up to 14 days storage at 5 °C. In a similar study, pectin edible coating solution containing pectin (3%), glycerol (2.5%), polyvinyl alcohol (1.25%), and citric acid (1%) was prepared and applied on sapota fruits by dipping method and uncoated sapota fruits were used as control. Both the treated and control fruits were stored at 30 ± 3 °C. Physicochemical parameters namely, weight, color, firmness, acidity, TSS, pH, and ascorbic acid contents of both the coated and control fruits were measured at regular interval up to 11th day of the storage at 30 ± 3 °C. Reduced rate of change in weight loss and other parameters were reported in pectin-coated sapota as compared to control fruits and it was observed that pectin film formulation was able to maintain good quality attributes and extend the shelf life of pectin-coated sapota fruits up to 11 days of storage at room temperature, whereas control fruits were edible up to 6 days. Furthermore, it was also observed that sapota fruits dipped in sodium alginate containing 2% pectin solution for 2 min were more effective in maintaining the organoleptic properties up to 30 days of refrigerated storage as compared to sapota fruits dipped for 4 min and untreated sapota fruits. Bayarri et al. developed antimicrobial films using lysozyme and LM pectin complex. The main purpose of the study was to control the release of lysozyme in packaged food and to target lysozyme-sensitive bacteria such as Bacillus and Clostridium. It was observed that in the presence of fungal pectinase, due to the dissociation of pectin linkage, lysozyme activity of films increased remarkably. Many food-contaminating bacteria are pectinase producing and such type of films may be used to control food contaminants. These results have opened new avenues for custom-made biodegradable film.

Edible coating formulation containing sodium alginate and pectin (PE) enriched with eugenol (Eug) and citral (Cit) essential oil at different concentrations was used to increase the shelf life of strawberries. Physical and organoleptic parameters of coated fruits stored at 10 °C for 14 days show that formulation containing PE 2% + Eug 0.1%; PE 2% + Cit 0.15% was more suitable than sodium alginate-based formulations. Pectin coating containing lemon and orange peel essential oils was reported to increase the shelf life and quality attributes of the strawberry fruits up to 12 days when stored at 5 °C. It was also observed that fruits coated with pectin + 1% orange essence showed less weight loss and soluble solids as compare to their control during the storage. Sanchís et al. studied the combined effect of edible pectin coating with active modified atmospheric packaging on fresh-cut "Rojo Brillante" persimmon. Persimmon fruit slices were coated by dipping in the pectin-based emulsion or in water as control. Both the treated and control slices were packed under 5 kPaO_2 (MAP) or under ambient atmosphere for up to 9 days at 5 °C. Various parameters, such as package gas composition, color and firmness of slice, polyphenol oxidase activity, were measured during storage. It was observed that edible coating along with MAP significantly reduced the CO_2 emission and O_2 consumption in the packaged fruits. Furthermore, coating was also effective in controlling microbial growth and reducing enzymatic browning and maintains good sensory parameters up to 10 days on storage.

Drying is the traditional and oldest method of fruit and vegetable preservation. It decreases the enzymatic activity, reduces the moisture contents, and protects the food from microbial attack. However, drying results in loss of nutrients, vitamins, heat-labile enzymes, modifies the texture, color, and organoleptic quality of dried fruits and vegetables and therefore diminishes the market value also. Pre-treatment of food products with pectin coatings containing other bioactive compound such as ascorbic acid, $CaCl_2$, edible gum, etc., before drying or blanching has been proposed as an effective method to preserve the nutritional as well as organoleptic quality of dried food. Recent researches have shown that application of pectin coating could protect the moisture and vitamin C loss in pre-treated papaya slice and osmotic dehydrated pineapple. Pineapple slice was pre-treated with pectin coating formulation containing (50%)/calcium lactate (4%)/ascorbic acid (2%) solutions and then dried by hot-air-drying method. Physicochemical analysis of dried product showed less reduction in vitamin C contents as compared to untreated pineapple slice. In a similar work, pectin coating supplement with vitamin C (1%) was used for pre-coating of papaya slice. It was found that incorporation of vitamin C did not affect the drying process. However, significant increase in vitamin C content was observed in final product.

Frying is a method of cooking that causes changes in chemical and physical parameters of food and enhances the taste. However, high temperature vaporizes the water of food and affects the nutritional properties due to protein denaturation and starch gelatinization. The oil uptake during frying is affected by various parameters such as type of oil used, frying temperature and duration, product moisture content, shape, porosity, pre-frying treatment, etc. Surface area and pre-treatment of products are the major factors that determine the oil absorbed. Edible coating has also been used successfully, to reduce the oil uptake during frying in various deep-fried products.

Cellulose

Cellulose is a molecule comprised of carbon, hydrogen, and oxygen, and is found in the cellular structure of virtually all plant matter. This organic compound, which is considered the most abundant on earth, is even excreted by some bacteria.

Cellulose provides structure and strength to the cell walls of plants and provides fiber in our diets. Although some animals, such as ruminants, can digest cellulose, humans cannot. Cellulose falls into the category of indigestible carbohydrates known as dietary fiber.

In recent years, cellulose has become a popular food additive due to its unique chemical and physical properties when combined with water. Although cellulose can be found in most plant matter, the most economical sources of industrial cellulose are cotton and wood pulp.

How Cellulose is Used in Food

Fiber Supplement: With rising awareness about fiber intake, cellulose has become one of the most popular food additives. Adding cellulose to food allows an increase in bulk and fiber content without a major impact on flavor. Since cellulose binds and mixes easily with water, it is often added to increase the fiber content of drinks and other liquid items when the gritty texture of regular fiber supplements would be undesirable.

Calorie Reducer: Cellulose provides a lot of volume or bulk of food but because it is indigestible to humans, it has no caloric value. For this reason, cellulose has become a popular bulking agent in diet foods. Consumers who eat foods with high cellulose content feel full physically and psychologically without having consumed many calories.

Thickening/Emulsifying: The gelling action of cellulose when combined with water provides both thickening and stabilizing qualities in the food to which it is added. Cellulose gel acts similarly to an emulsion, suspending ingredients within a solution and preventing water from separating out. Cellulose is often added to sauces for both the thickening and emulsifying action.

The thickening power of cellulose also allows for more air to be whipped into products like ice cream, or whipped topping. Cellulose allows for the production of thick and creamy food items without the use of as much fat.

Anti-caking: Cellulose's ability to absorb moisture and coat ingredients in a fine powder makes it the ingredient of choice for anti-caking applications. Shredded and grated cheeses, spice mixes, and powdered drink mixes are just a few of the many food items that take advantage of cellulose as an anti-caking agent.

Forms of Cellulose

Cellulose can be found on ingredient lists under a variety of names, depending on which form is used. Although cellulose has the same molecular structure regardless of the source (wood pulp, cotton, or other vegetable matter), how the molecules are bonded together and whether or not they are hydrated creates different "forms" of cellulose.

Powdered cellulose is the most widely used in food products and is the form of choice for anti-caking applications. Cellulose gum or cellulose gel, which are hydrated forms of cellulose, are often used in sauces or other wet items like ice cream and frozen yogurt. Cellulose may also be found on ingredient lists under the names carboxymethylcellulose, microcrystalline cellulose, or MCC.

Starch

Starch is an important food product and a versatile biomaterial used world-wide for different purposes in many industrial sectors including foods, health, textile, chemical and engineering sector. Starch also known as amylum, is an important food product and biomaterial used world-wide for different purposes. Though traditionally used in the food industry, technological advancement has led to its steady relevance in many other sectors such as health and medicine, textile, paper, fine chemicals, petroleum engineering, agriculture, and construction engineering. It is used in the food industry either as food products or additives for thickening, preservation and quality enhancer in baked foods, confectioneries, pastas, soups and sauces, and mayonnaises. Starch is a polysaccharide of glucose made of two types of α-d-glucan chains, amylose and amylopectin. Starch molecules produced by each plant species have specific structures and compositions (such as length of glucose chains or the amylose/amylopectin ratio), and the protein and fat content of the storage organs may vary significantly. Therefore, starch differs depending on the source. This inherent functional diversity due to the different biological sources enlarges its range of industrial uses.

The structural and compositional differences in starches from different sources determine its properties and mode of interactions with other constituents of foods that gives the final product the desired taste and texture. In the food industry, starch can be used as a food additive to control the uniformity, stability and texture of soups and sauces, to resist the gel breakdown during processing and to raise the shelf life of products. Starch is relatively easily extractable and does not require complicated purification processes. It is considered to be available in large quantities in major plant sources such as cereal grains and tubers. These sources are generally considered inexpensive and affordable and serve as raw materials for commercial production.

Starch from Zea mays (corn) account for 80% of the world market production of starch. Maize starch is an important ingredient in the production of many food products, and has been widely used as a thickener, stabiliser, colloidal gelling agent, water retention agent and as an adhesive due to its very adaptive physicochemical characteristics. Starches from tubers of roots such as potato tubers, which are considered non-conventional sources have found usefulness in providing options for extending the spectrum of desired functional properties, which are needed for added-value food product development.

(A) Corn and (B) potato tuber.

The stability of native starch under different pH values and temperatures varies un-favorably. For instance, native starch granule is insoluble in water at room temperature and extremely resistant to hydrolysis by amylase. Hence native starch has limited functionality. In order to enhance properties and functionality such as solubility, texture, viscosity and thermal stability, which are necessary for the desired product or role in the industry, native starches are modified. The widening vista of application possibilities of starches with different properties has made research in non-conventional starches and other native starches more imperative. Recent studies on the relationship between the structural characteristics and functional properties of starches from different sources have continued to provide important information for optimizing industrial applications.

Modification has been achieved mostly by physical and chemical means. Enzymic and genetic modifications are biotechnological processes which are increasingly being explored. While physical modification methods seemed simple and cheap, such as superheating, dry heating, osmotic pressure treatment, multiple deep freezing and thawing, instantaneous controlled pressure-drop process, stirring ball milling, vacuum ball milling, pulsed electric fields treatment, corona electrical discharges, etc., chemical modification involves the introduction of new functional moieties into the starch molecule via its hydroxyl groups, resulting in marked change in its physicochemical characteristic. The functional characteristics of chemically modified starch depends on a number of factors including the botanic origin of the native starch, reagent used, concentration of reagent, pH, reaction time, the presence of catalyst, type of substituent, degree of substitution, and the distribution of the substituents in the modified starch molecule. Modification is generally achieved through chemical derivatization, such as etherification, esterification, acetylation, cationization, oxidation, hydrolysis, and cross-linking.

Amylose and Amylopectin

The chemical behaviour of starch is dependent on the nature of its constituent compounds. Starch is a homopolysaccharides made up of glucose units. However, the homopolysaccharide are of two types namely: amylose, which is a linear chain consisting of about 500–2000 glucose units, and amylopectin, which is highly branched and consist of over 1,000,000 glucose units. The two types of homopolysaccharides constitute approximately 98–99% of the dry weight of starch. The ratio of the two polysaccharides usually varies depending on the botanical origin of the starch. Botanic source reports that starch chain generally consist of 20% amylose and up to 80% amylopectin by mass. It is believed that starch with up to 80% amylose can exist. Some classification categorize starch containing <15% amylose as 'waxy', 20–35% as 'normal' and greater ≥40% as 'high' amylose starches.

Amylose and amylopectin have different physiochemical properties which impact on the overall properties of the starch. Hence it is often important to determine the concentration of each individual component of the starch, as well as the overall starch concentration. The physicochemical (e.g., gelatinization and retrogradation) and functional (e.g., solubility, swelling, water absorption, syneresis and rheological behaviour of gels) properties determine the potential uses of starches in the food industry. These properties depend on the molecular and structural composition of amylose and amylopectin, percent composition and arrangement of these two homopolysaccharides in starch granules which often determine the granule size and shape depending on other genetic factors as a result of the particular species of plant.

In food products, the functional roles of starch could be as a thickener, binding agent, emulsifier, clouding agent or gelling agent. In the food industry, native starch is usually reprocessed and modified through chemical processes to improve its functionality for the desired purpose. Chemical modification involves the introduction of new functional groups into the starch molecule which produces in a modified starch with markedly altered physicochemical properties. Such modified starch shows profound change in functionality such as solubility, gelatinization, pasting and retrogradation.

The chemical reactivity of starch is dependent on the reactivity of the constituent glucose units. The chemical and functional properties achieved following such modification depends largely on the reaction conditions such as modifying reagent(s), concentration of the reactants, reaction time, type of catalyst used, pH, and temperature. The type of substituents, degree of substitution and distribution of substituents in the starch molecule affects the functional properties.

Amylose

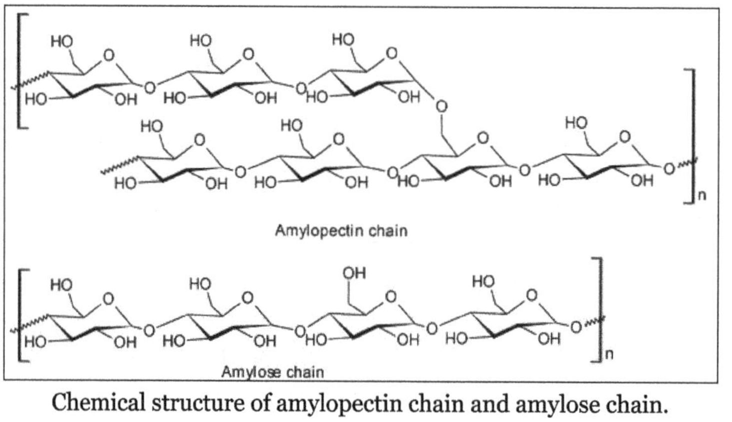

Chemical structure of amylopectin chain and amylose chain.

Amylose is a linear polymer of α-d-glucose units linked by α-1, 4 glycosidic bonds. The linear nature of amylose chain and its percentage content in starch, and the relative molecular arrangement with amylopectin affect the overall functionality of the starch. Hence starch varies greatly in form and functionality between and within botanical species and even from the same plant cultivar grown under different conditions. This variability provides starches of different properties, which can create challenges of raw materials inconsistency during processing.

Amylopectin

Amylopectin is a branched polymer of α-d-glucose units linked by α-1, 4 and α-1,6 glycosidic bonds. The α-1, 6 glycosidic linkages occurs at the branching point while the linear portions within a branch are linked by α-1, 4 glycosidic bonds. In comparison to amylose, amylopectin is a much larger molecule with a higher molecular weight and a heavily branched structure built from about 95% (α-1,4) and 5% (α-1,6) linkages. Amylopectin unit chains are relatively short with a broad distribution profile, compared to amylose molecules. They are typically, 18–25 units long on average.

Physicochemical Properties of Starch

Physical properties are those properties exhibited without any change in chemical characteristics of starch and do not involve the breaking and creation of chemical bonds such as solubility, gelatinization, retrogradation, glass transition, etc. On the other hand, chemical properties changes due to chemical reactions and usually involve the breakage and creation of new bonds. Examples of such chemical processes in starch include hydrolysis, oxidation, esterification and etherification. Research strongly indicates that the molecular weight and branching attributes of starch which play important roles in the shape and size of granules can potentially be used for predicting some of its functionality such as texture, pasting, retrogradation, etc.. Amylose has more proportional relationships with pasting and gel textural properties, while amylopectin which are predominant in regular and waxy corn starches, has higher proportional relationship with firmness.

Solubility and Gelatinization

When unprocessed or native starch granules which are relatively inert are heated in the presence of adequate water, usually during industrial processes, swelling of the granules occur and the amylose dissolves and diffuses out of the swollen granules which upon cooling forms a homogenous gel phase of amylose-amylopectin. The swollen amylopectin-enriched granules aggregate into gel particles, generating a viscous solution. This two-phase structure, called starch paste, is desirable for many food applications where processed starches are used as thickeners or binders.

Retrogradation and Shear

Retrogradation of starch is a phenomenon that occurs when the disordered arrangement of the polymer molecules of gelatinized starch begins to re-align into an ordered structure in the food product. Preventing retrogradation affects the freeze-thaw stability and textural characteristics and helps to elongate the shelf life of the food product. Starch modification through chemical means, such as, hydrolysis and esterification are generally used to produce starches that can withstand retrogradation. Preventing retrogradation of starch is important for starch used in frozen foods because it is accelerated at cold temperatures, producing an opaque, crystallized, coarse texture as a result of the separation of the liquid from the gel or syneresis. Crosslinked oxidized starches have been reported be more stable against retrogradation.

Amylose linear chain dissolves in water at 120–150 °C and is characterized by high thermostability, resistance to amylase, high crystallinity and high susceptibility to retrogradation. Amylopectin, which is the branched chain is however, slow to retrogradation, with crystalline forms appearing only on the outside of the globule and characterized by a significantly lower re-pasting temperature of 40–70 °C and an increased susceptibility to amylases activity than amylose. Retrogradation of starch is affected botanical origin of the starch, amylose content, length of the amylopectin chains, density of the paste, paste storage conditions, physical or chemical modifications and the presence of other compounds. Recrystallization of starch applies only to amylose chains, and it occurs most readily at temperatures around 0 °C, and also at temperatures above 100 °C. Physical modification process such as repeated freezing and thawing of the starch paste aggravate retrogradation. The resulting starch thus produced is resistant starch that exhibit resistance to digestibility by amylase enzymes and can be used as an alternative nutrient source for diabetic patients and as a rate controlling polymer coat in controlled drug delivery systems.

Starch granules swollen with water are predisposed to fragmentation if exposed to physical severe pressure change. This becomes of major concern where the integrity of the granules is required to maintain viscosity. Shear is the disintegration phenomenon of swollen starch granules or gel. Starch shear arises from the shear stress which builds up during the process of retrogradation and/or gel drying of the gelatinized starch. The stress acting in opposite directions creates a fault-line that causes the material to open up or tear apart. Shearing generally depends on the fluid (gel) viscosity and flow velocity. Starch granules in their raw unswollen forms are not susceptible to damage by shear even in the slurry before cooking. But once cooked or gelatinized, starch granules becomes susceptible to shear, resulting in loss of viscosity and textural stability.

Chemical Properties of Starch

The chemical properties of starch are dependent on the reactivity of starch which is a function of the polyhydroxyl functional groups in the constituent glucose monomers. The hydroxyl groups at position C-2, C-3 and C-6 which are free from the glycosidic bond linkages and pyranose ring formation, are usually free for substitution reactions involving either the attached hydrogen or the entire hydroxyl group. While the —OH at C-6 is a primary alcoholic hydroxyl group, those at C-2 and C-3 are secondary alcoholic hydroxyl group. Hence starch can undergo hydrolytic cleavage of its chains at the glycosidic bonds; oxidative reaction with the —OH or C—C bond creating carbonyl groups; and other reactions with various functional and multifunctional reagents to produce esterified and etherified starches. Most of the reactions require activation of the hydroxyl of glucose units in acidic or basic media.

Reactions of Starch

The reactivity of starch is dependent on the hydroxyl functions of the constituent α-D-glucan polymers. Thus starch is able to undergo the following reactions.

Hydrolysis

Hydrolysis is an addition reaction and simply involves the addition of a water molecule across a bond resulting in the cleavage of that bond and formation of the cleavage products, usually with hydroxyl group or alcohol functionality. Hydrolysis of starch can be achieved by chemical or enzymatic process. Chemical process of hydrolysis usually employs heating starch in the presence of water or dilute hydrochloric acid. Hydrolysis is also used to remove fatty substances associated with native starches. Hydrolysis under acidic condition is called roasting, resulting in acid modified starch. Treatment of starch with sodium or potassium hydroxide results in alkaline modified starch. Hot aqueous alkaline solutions can be used, and this improves the reducing value of that starch.

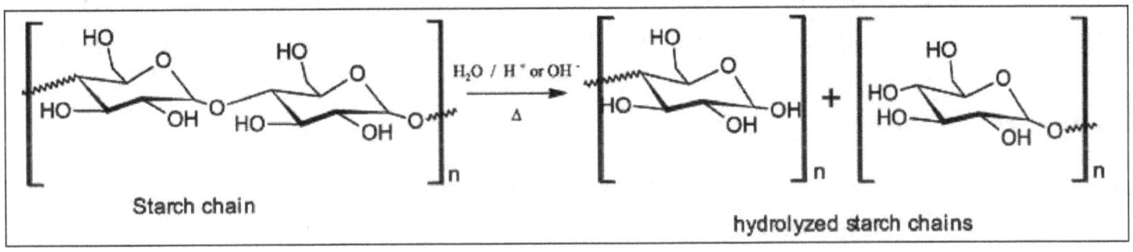

Hydrolysis of α(1 → 4) glycosidic bond.

The products of starch hydrolysis include dextrin or maltodextrin, maltose and glucose. Dextrins are mixtures of polymers of d-glucose units linked by α-(1 → 4) or α-(1 → 6) glycosidic bonds. The percentage of products obtained depends on the conditions used for the reaction such as duration and strength/amount of reagents used. Enzymic hydrolysis uses the enzyme malto-amylase to achieve hydrolysis and this is the process that usually occurs in starch digestion in the gastrointestinal tract. Dextrins are white, yellow, or brown water-soluble powder which yield optically active solutions of low viscosity. Most of them can be detected with iodine solution, giving a red coloration. White and yellow dextrins from starch roasted with little or no acid are called British gum. The properties of dextrinized starch is dependent upon the reaction conditions (moisture, temperature, pH, reaction time) and the products characteristics vary in its content of reducing sugar, cold water solubility, viscosity, color and stability.

Hydrolytic processes have been used in the food industry to produce starch derivatives with better functional properties and processing applications. Acid and alkali steeping are the two most widely used methods for starch isolation in the food industry, with numerous modifications. Thermo-alkali isolation method known as nixtamalization has been used in Central America since pre-Hispanic times. Acid and alkali isolation processes affect the amylose/amylopectin, protein and lipid content as well as the granule size and shape of the final product.

Esterification Reaction

The condensation of an alcohol and carboxylic acid usually under acidic condition, to produce an ester and water, is called esterification. Basically, the reaction is between the carboxylic acid group and the alcohol group with the elimination of a water molecule. When the acid anhydride is used, an alkaline condition is preferred in the reaction.

$$RCOOH + R'\text{-}OH \rightleftharpoons RCOOR' + H_2O$$

Esterification reaction of carboxylic acids and alcohols.

The reaction is usually reversible and the forward reaction is favoured under low pH and excess of alcohol while the reverse is favoured under high pH. Remover of one of the product during the reaction will also favour the forward reaction.

For starch, the reaction is between the carboxylic acid group (—COOH) of fatty acids or —COCl of fatty acid chlorides and the alcohol group (—OH) of the glucose units. Esterification is generally used to introduce more lipophilic groups into the starch molecule making it more lipophilic and for producing crosslink starch when polyfunctional compounds or multifunctional or reagents capable of esterification or etherification are used. Esterification weakens the inter-molecular bonding that holds the granules together and hence alter the granule shape and sizes as well as other functional properties of the starch. The degree of substitution (DS) is dependent on the concentration of reagent used, the type of reagent used, the catalyst and the duration of reaction.

Acetylation of Starch

Starch can be acetylated by reacting it with acetic anhydride to produce acetylated starch. The hydroxyl group of the glucose units are esterified with the acetyl groups from the acetic anhydride to give starch with glucose units with acetate function. The DS of the hydroxyl group with acetate group is dependent on the reaction conditions. Acetylated corn starch of DS 0.05, 0.07 and 0.08 have been obtained using 4, 6 and 8% (starch d.w.) acetic anhydride respectively and aqueous sodium hydroxide as catalyst.

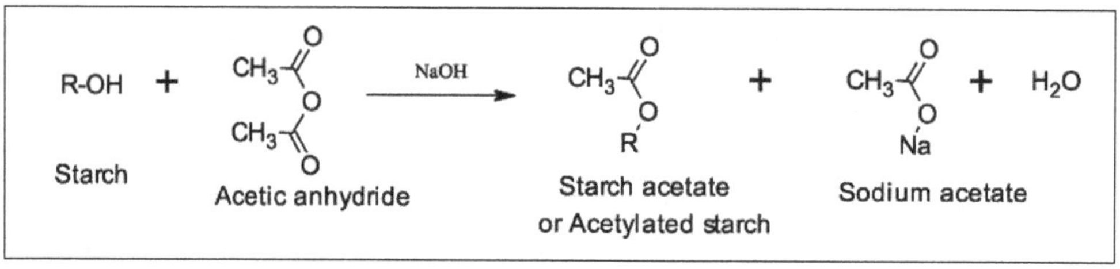

Acetylation of starch with acetic anhydride.

The introduction of the more bulky acetyl group compares with hydroxyl group causes steric hindrance to the alignment of the linear chains. This allows for easy water percolation between chains thus increasing the granule swelling power and solubility resulting in lower gelatinization temperature. The steric hindrance of less polar acetyl group also reduces the amount of inter-molecular hydrogen bond formation, and weakens the granule structure, preventing molecular re-association and realignment required for retrogradation. However, depending on the DS and the interplay between the a weakened granular structure as result of interruption of the inter- and intra-molecular bonds, and reduced bonding with water molecules as a result of the hydrophobicity of the acetyl groups, the viscosity of the final product can be enhanced.

Acetylation improves paste clarity and freeze-thaw stability of starch. Starch acetates of low DS are commonly used in the food industry for quality consistency, and as texture and stability enhancers. The Food and Drug Administration (FDA) maximum DS of acetylated starches for food application is 0.1. Starch acetate of high DS exhibit high degree of hydrophobicity and thermo-plasticity and are soluble in organic solvents like chloroform and acetone, and are mostly used in non-food applications. At 0.0275 DS, corn starch exhibit lower paste gelling, which is practically lost at 0.05 DS. Most commercial starch acetates have <0.05 DS.

Acetylated distarch adipate, is a monosubstituted starch obtained by treating starch with acetic anhydride and adipic anhydride. It has been used since the 1950s due to desire for improved stability of product in cold and freezing weather conditions. It is a good temperature change resistant agent used in foods as a bulking agent, stabilizer and thickener. It improves smoothness and sheen of soups and sauces. The improved freeze-thaw stability of acetylated cross-linked waxy maize starch has led to its use in frozen sauces in vegetables, appetizers and pastries. Hydroxypropylation of cross-linked starch also dramatically improves the stability quality of puddings and frozen sauces.

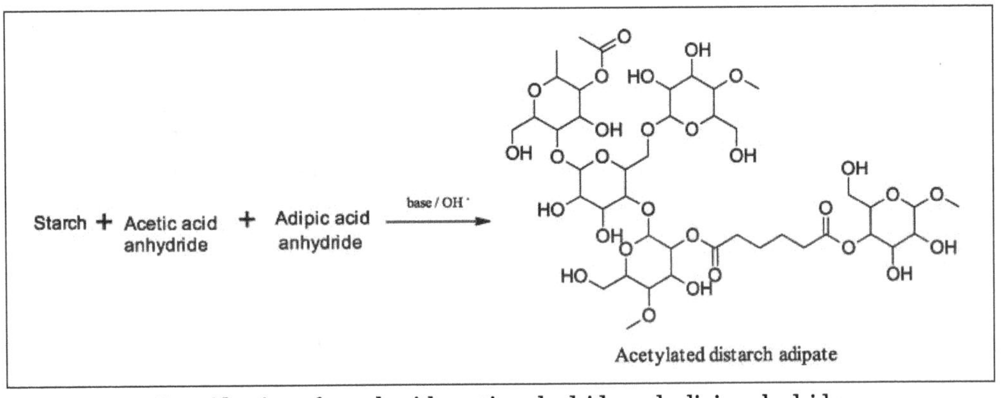

Esterification of starch with acetic anhydride and adipic anhydride.

Succinylation of Starch

When starch granule is esterified with succinic anhydride, it produces succinyl starch, and the process is commonly referred to as succinylation of starch. Succinylation of starch was earlier achieved in the presence of aqueous pyridine and under reflux at 115 °C. However, environmental concerns have led to the development of more green synthetic routes. Thus succinic ester of starch have been prepared by mixing starch with succinic anhydride solution in acetone and refluxing at 110 °C for 4 h. Sui et al. was also able to induce a reaction by drop-wise addition of succinic anhydride to a water suspension of starch while maintaining pH at 8.5 by drop-wise addition of sodium hydroxide.

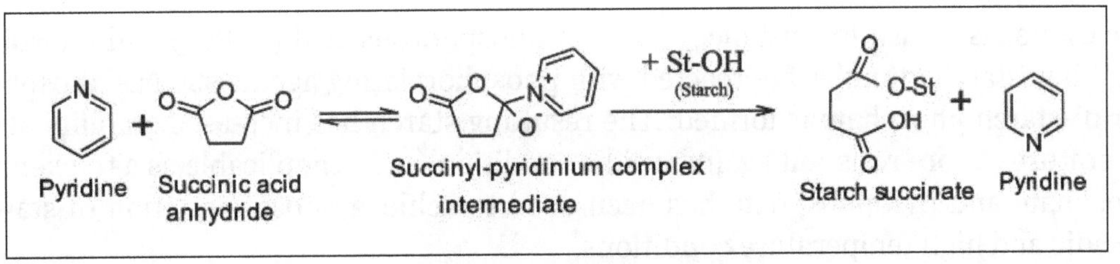

Succinylation reaction of starch.

Succinyl group weakens the inter-molecular bonding of starch polymeric chains in the granules, facilitating swelling, solubilisation and gelatinization at lower temperatures. Paste clarity is enhanced and retrogradation is reduced. However, there may be reduced stability against shear at high temperature and during cooling. Starch succinate is ionic and acts as polyelectrolytes. At low degree of substitution (DS), the succinate makes the starch more hydrophilic and viscos in solution. For its viscosity enhancing effect, succinylated starches could find application in production of non-gelling custard creams, and for its increased hydrophilicity, it could be used for enhancing the juicy/smooth taste of meat and fried products. Starch succinates can also be used in soups, snacks, and frozen/refrigerated food products as thickening or stabilizing agents.

Esterification of starch with octenylsuccinic anhydride (OSA) or octenylsuccinic acid in the presence of an alkali yields starch octenylsuccinate, while esterification with dodecyl succinic acid yield starch dodecyl succinate. The octenyl or dodecyl group introduce a reasonable level of lipophilicity to the product making it have dual functionality which can be used in emulsification and flavours encapsulation. OSA treated starches are used to stabilize oil-in-water food emulsions associated with beverage concentrates containing flavor and clouding oils. It helps to protect emulsified and spray dried flavour oils against oxidation during storage. FDA allows a DS of 0.02.

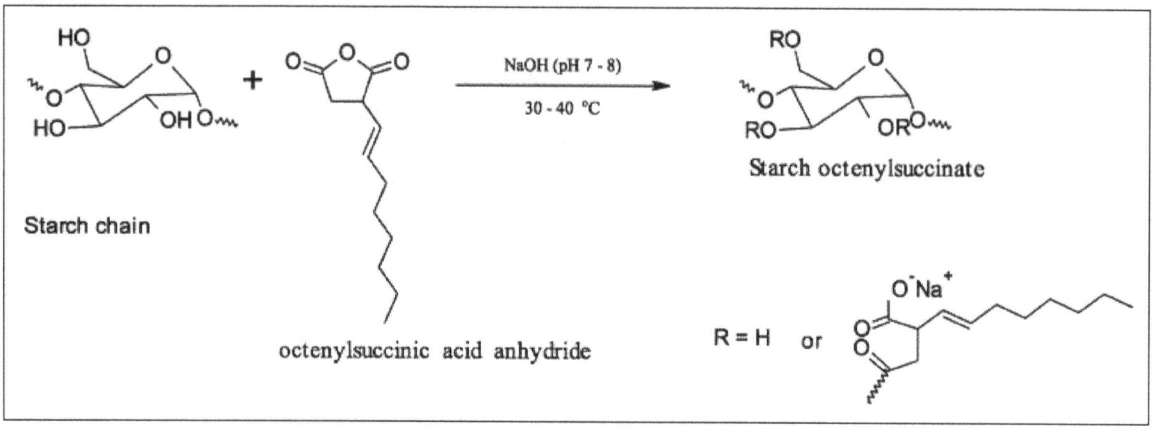

Esterification of starch with octenylsuccinic acid anhydride.

Commercial production of acetylated starch dodecyl succinate, di-substituted starch of low dodecyl succinate residue employs acetic anhydride reagent at alkaline pH. An alkali-starch complex forms first, which then interacts with the carboxylic anhydride to form a starch ester with the elimination of carboxylate ion and one molecule of water. Starch succinate offers freeze-thaw stability, high-thickening, low-gelatinization temperature, clarity of paste, good film-forming properties and resistance to retrogradation.

Phosphorylation Reaction

Inorganic esters also exist, for instance, esters of phosphorous acid (H_3PO_3) and phosphoric acid (H_3PO_4). When starch granules are reacted with phosphorylating agents such as phosphoric acid, mono- or di-starch phosphate is formed. The resulting starch has increased stability at high and low temperatures, more resistant against acidic condition, and is applicable as a thickening agent. Orthophosphate and pyrophosphate has been used to achieve phosphorylation of starch under slightly acidic and high temperature conditions.

$$\underset{\text{Starch}}{\text{St-OH}} + \underset{\text{Pyrophosphoric acid}}{H_4P_2O_7} \xrightarrow{\text{pH 3-6/ heat}} \underset{\text{Starch phosphate}}{\text{St-O-}H_2PO_3} \quad \underset{\text{Phosphoric acid}}{H_3PO_4}$$

Phosphorylation reaction of starch.

Phosphoryl trichloride, sodium tripolyphosphate and sodium trimetaphosphate have also been used under higher pH to obtain monostarch phosphate and di-starch phosphate. Phosphorylation reactions produce either monostarch phosphate or distarch phosphate which is a cross-linked derivative. However this depends on the reagents and reaction conditions. Usually, monoesters, rather than diesters, are produced with a higher degree of substitution. Steric hindrance as a result of the introduced phosphate groups inhibits the linearity of amylose or the outer branch of the amylopectin chain where it reacted. This weakens the inter-molecular association and creates chains disaggregation, which leads to better paste clarity.

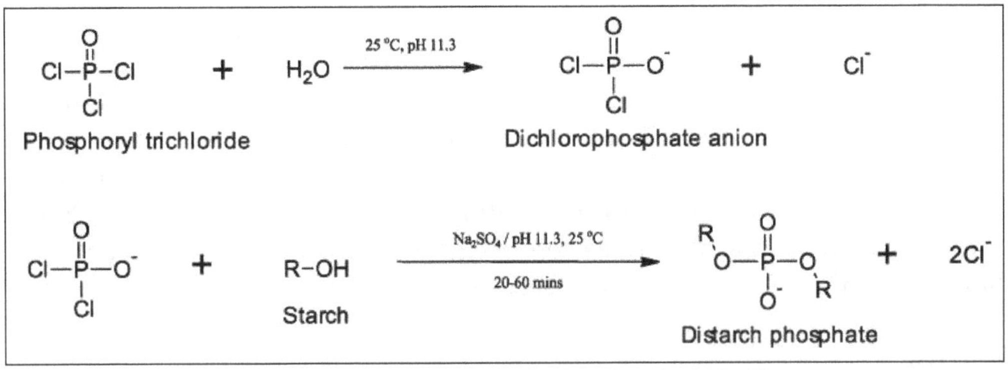

Phosphorylation of starch with phosphoryl trichloride.

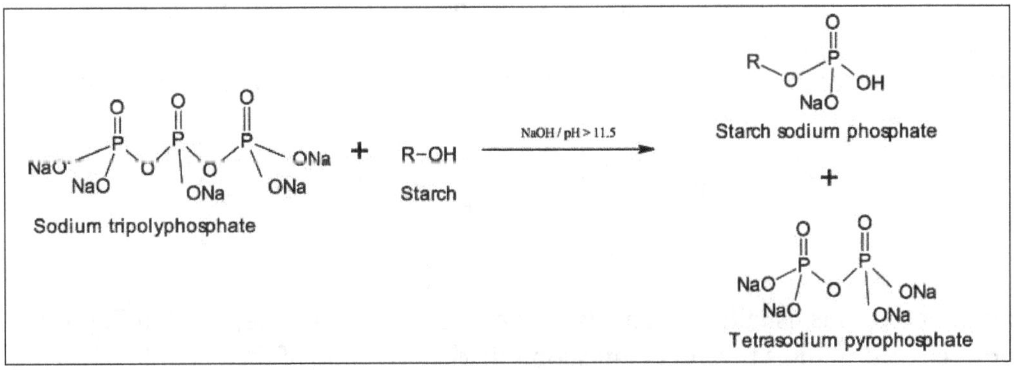

Phosphorylation of starch with sodium tripolyphosphate.

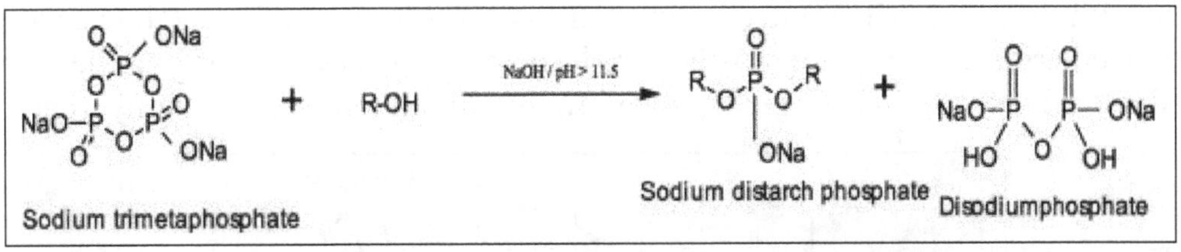

Phosphorylation of starch with sodium trimetaphosphate.

Distarch phosphate has the phosphate group esterified with two hydroxyl groups of two neighbouring starch polymer chains. The phosphate bridge or cross-linking strengthens the mechanical structure of the starch granules. Phosphate cross-linked starches exhibit stability against high temperature, low pH and shear, and improved firmness of the swollen starch granule as well as improved viscosity and textural characteristic. Distarch phosphate is used as thickener and stabilizer and provides stability against gelling and retrogradation and high resistance to syneresis during storage.

In solution, several specie of the phosphate ion can exist and anyone may be responsible for the phosphorylation reaction depending on the reaction conditions. Phosphorylation has been demonstrated to mostly occur at the C-3 and C-6 of the glucose units, and the degree of phosphorylation depends on distribution of the chain length of the starch polymers. Blennow et al. also demonstrated that phosphate groups may play important role in the size distribution of the amylopectin side chains of phosphorylated starches. Some researchers have reported that about 60–70% of total phosphorus of starch monophosphate is located at C-6 while the rest is located at C-3 of anhydroglucose units. Most phosphate groups (88%) are on chain β of amylopectin.

Phosphorylated starch prepared by the slurry treatment exhibited a lower gelatinization temperature, a higher peak viscosity, a lesser degree of retrogradation, and improved freeze-thaw stability compared with those prepared by the dry-mixing treatment. They believed that phosphorylation probably occurred in both amylose and amylopectin chains, and the amount and location of incorporated phosphate groups varied with starch types, which may be due to their different amylose and amylopectin contents. Waxy starch was more prone to phosphorylation, followed by common and high-amylose starches. Enzymic phosphorylation of starch has been reported. Extrusion condition of 200 °C, sodium tripolyphosphate concentration of ≥1.4 g/100 ml and pH 8.5 have been used to obtain starch phosphate with high degree of substitution.

Etherification

Generally, alcohols (—OH) groups condenses with one another at high temperatures under acidic conditions to form ethers. The reaction mechanism is through a proton transfer from the catalyst to one of the molecule to form a cation, which loses the proton by extracting the —OH of the second molecule to form an ether and water.

$$R^1\text{-OH} + R^2\text{-OH} \xrightarrow[H^+/\text{heat}]{} \underset{\text{Ether}}{R^1\text{-O-}R^2} + H_2O$$

Etherification reaction.

Etherification of starch is usually done by use of epoxide reagents as depicted in Figures. The epoxides are first reduced to diols through a nucleophilic ring opening of the epoxide (cleaving the C—O bond under aqueous, acidic or alcoholic condition) before the eventual condensation of one of the —OH group with that of starch. Some etherification reactions occur under alkaline condition. Like esterification, etherification helps to mostly introduce lipophilic alkyl groups into the starch chains thereby reducing the hydrophilicity and the degree of inter- and intra-molecular hydrogen bonding.

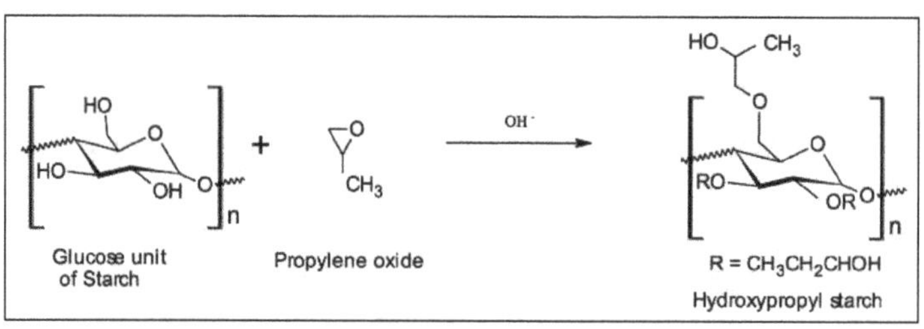

Etherification of starch with propylene oxide.

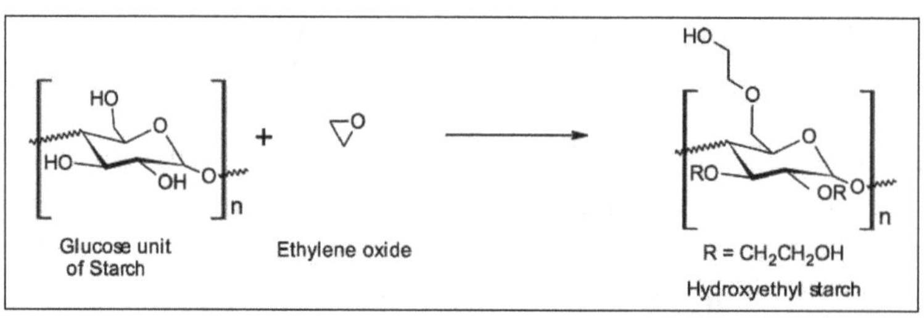

Etherification of starch with ethylene oxide.

Hydroxypropylation of Starch

This reaction process produces hydroxypropylated starch (HPS), which is a starch ether produce by reaction of starch with propylene epoxide in the presence of an alkaline catalyst. HPS is used for enhancing stability and viscosity of food products. The hydroxypropyl groups introduced into the starch chains affect the inter- and intra-molecular hydrogen bonds, thereby allowing for more ease of displacement of starch chains in the amorphous regions. HPS is more stable to prolonged high temperatures than starch acetate especially at pH 6, and has improves freeze-thaw stability. It is mostly used in refrigerated or frozen foods and in the dairy industry. The FDA allowable DS for HPS is 0.2.

Hydroxyethylation of Starch

Hydroxyethylation of starch is performed by reacting starch with epoxyethane or ethylene oxide to produce the starch ether, hydroxyethylated starch (HES). The health concerns of hydroxyethylated starch are limiting its use in the food industry. However they are mostly used in medicine and pharmaceuticals as plasma volume expander and extracorporeal perfusion fluids.

Carboxymethylation of Starch

This is an etherification reaction process where starch is reacted with sodium chloroacetate or chloroacetic acid under certain conditions to produce carboxymethylated starch (CMS). The reaction involves refluxing chloroacetic acetic acid with dry starch (anhydroglucose units) in the presence of sodium hydroxide in a solvent mixture of ethanol/isopropanol (ratio 3:5). Anhydroglucose unit can be obtained from acid hydrolysed starch.

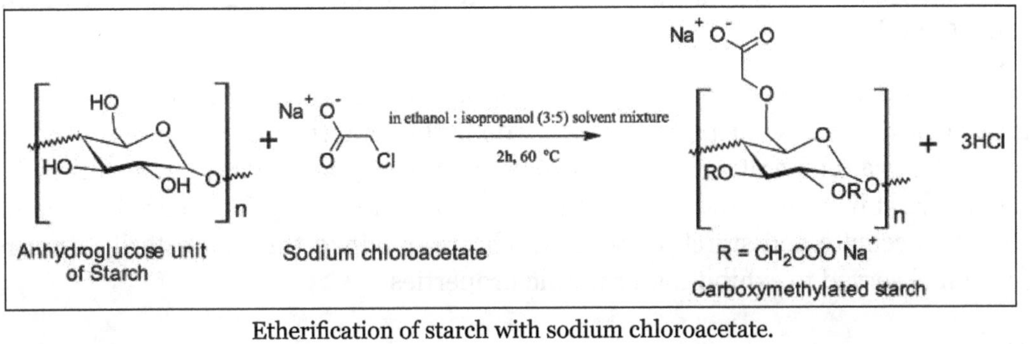

Etherification of starch with sodium chloroacetate.

Cationization of Starch

Another etherification reaction is cationization of starch in which starch react with electrophiles or electron-withdrawing reagents such as ammonium, amino, imino, sulfonium, or phosphonium groups to produce cationic starches, which are important industrial derivatives. Cationic starches are usually prepared under alkaline conditions, and they exhibit higher dispersibility and solubility with better transparency and stability.

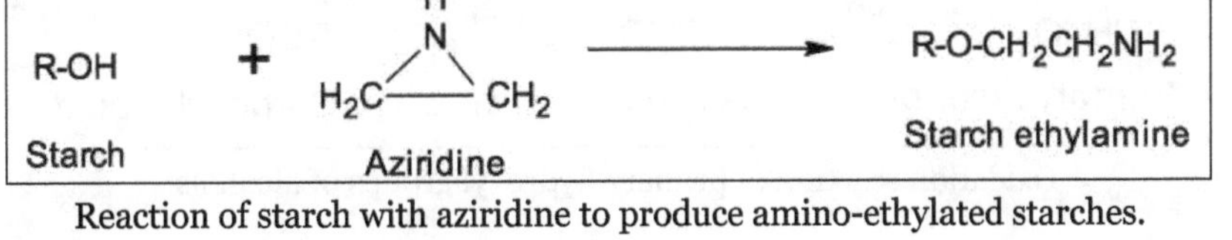

Reaction of starch with aziridine to produce amino-ethylated starches.

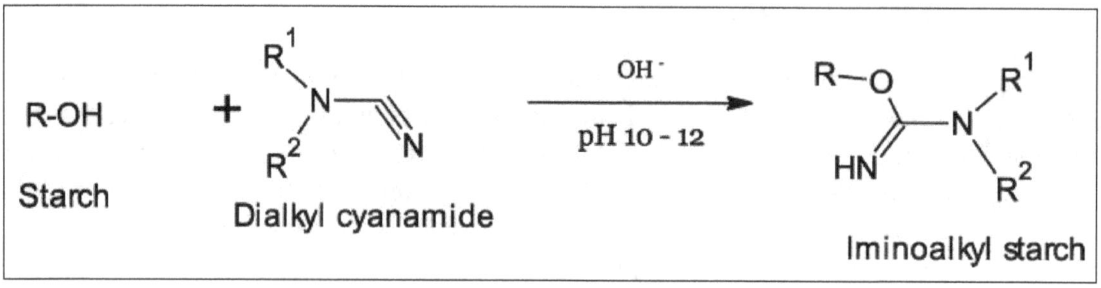

Reaction of starch and dialkyl cyanamides to produce aminoalkyl starches.

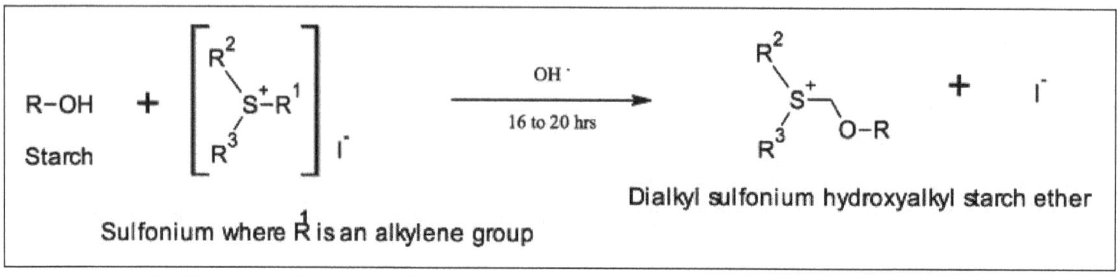

Etherification of starch with sulfonium salt to produce a sulfonium cationic starch.

For the production of sulfonium starch, halogenoalkyl sulfonium salts (e.g., 2-chloroethyl-methyl-ethyl sulfonium iodide or any β-halogenoalkyl sulfonium salt), vinyl sulfonium salts and the epoxy alkyl sulfonium can be used. Usually R^1 is unsaturated group like alkylene, hydroxyalkylene, aralkylene, cycloalkylene, and phenylene group, while each of R^2 and R^3 can be alkyl, aryl, aralkyl, cycloalkyl and alkylene sulfonium groups and may also contain ether oxygen linkages and amino groups. Factors such as reagent used and temperature, affect the reaction period which usually takes about 16–20 h.

Sulfonium starch display positive charge and can be used as thickeners in the form of aqueous dispersions or pastes. These dispersions are made by heating the suitable amount of sulfonium starch and water to a temperature of approximately 93 °C. Upon cooling, the resulting dispersion becomes considerably clearer and more resistant to viscosity change compared to the untreated starch. Starch succinate and starch citrates which are obtained through esterification reactions have also been observed to exhibit high cationic properties.

Oxidation

Oxidation of starch with strong oxidizing agents mimics reaction of primary alcohols and diols. Primary alcohol —OH functions are oxidized to its corresponding carbonyls (aldehydes and carboxylic acid), while vicinal diols are cleaved by strong oxidants like periodic acid into its corresponding carbonyl compounds (aldehyde and/or ketones). Oxidation of secondary alcohol —OH produces ketones. Oxidation may result in breakage of some intra- and inter-molecular bonds and partial depolymerization of the starch chains.

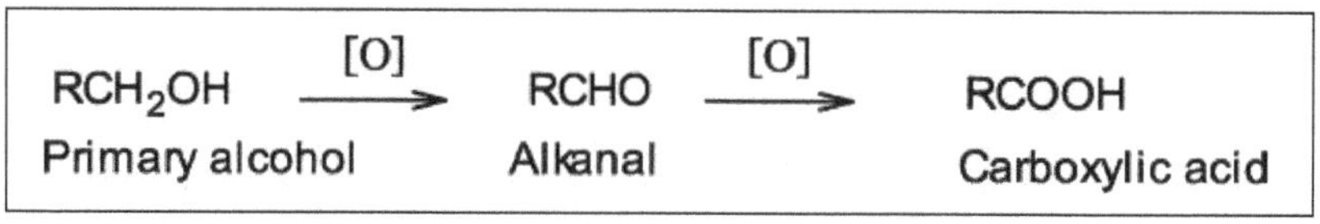

Oxidation reaction of primary hydroxyl groups of alcohols.

$$R^2CH(OH)CH(OH)R^1 \longrightarrow R^2C{=}O + O{=}CR^1$$

Oxidation reaction of vicinal hydroxyl groups of alcohols.

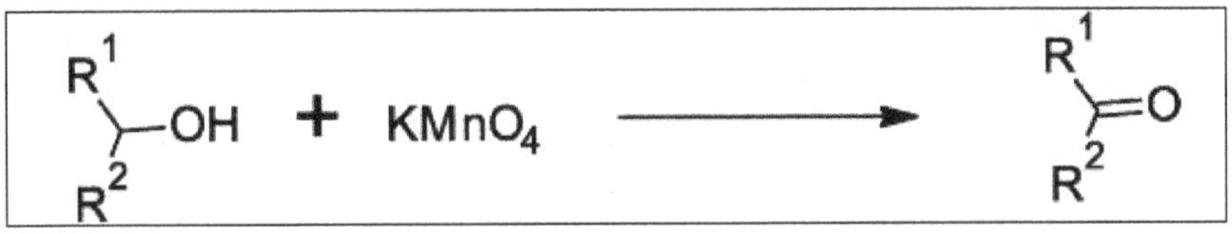

Oxidation reaction of secondary hydroxyl groups of alcohols.

Starches treated with oxidants fall into two broad classes: oxidized and bleached. Oxidized starches are starches treated with oxidizing agents like sodium hypochlorite (NaOCl). The oxidizing agent can attack the glycosidic bonds hydrolysing them to alcohol (–OH) functions or/and C–C bonds of the glucose unit, oxidizing them to carbonyl functions of aldehydes, ketones and carboxylates. Higher pH favors formation of carboxylate groups over aldehydes and ketones. Some depolymerization usually occurs in the process. Introduction of carboxylate groups provides both steric hindrance and electrostatic repulsion. Oxidation is usually carried out on whole granules and it causes the granule to dissolve, rather than swell and thicken. The reaction can introduce up to 1.1% of carboxyl groups in the granule. Oxidation with chlorine or sodium hypochlorite reduces the tendency of amylose to associate or retrograde. The reaction rate of starch with hypochlorite is remarkably affected by pH, which tend to be higher at about pH 7 but becomes very slow at pH 10. Oxidized starches are used where intermediate viscosity and soft gels are desired, and where the instability of acid-converted starches is unacceptable. Hence, pastes of oxidized starches have a lower tendency to gel compared to those of thin-boiling (or acid hydrolized) starches of comparable viscosity.

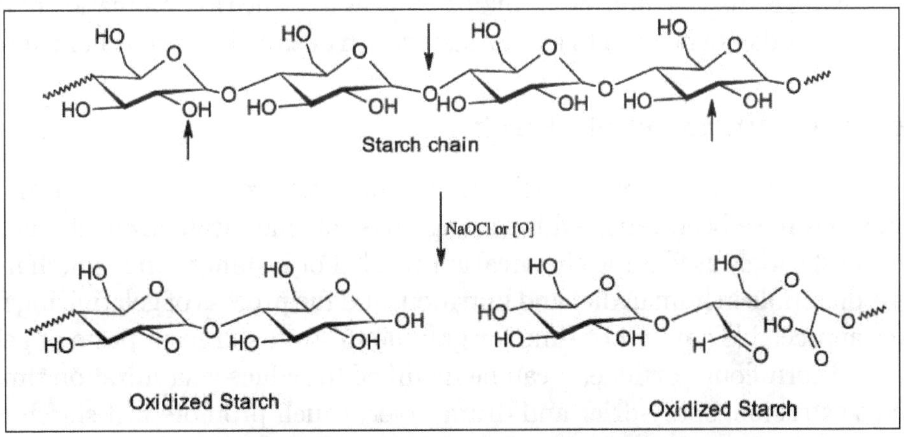

Oxidation reaction of starch to produce oxidized starch.

Other oxidants such as chlorine, hydrogen peroxide and potassium permanganate, dichromates and chlorochromates, etc. are less commonly used. Oxidized starches are reported to give batters improved adhesion to meat products and are widely used in dough and baked foods.

Bleached starch is obtained from oxidation of starch with lower concentrations of oxidizing agents like hydrogen peroxide, sodium hypochlorite, potassium permanganate or other oxidants used to remove color from naturally occurring pigments. Bleaching is done to improve the whiteness and/or eliminate microbial contamination. Reagent levels of about 0.5% are usually used, and loss of some starch viscosity due to hydrolysis usually occurs.

Cross-linking of Starch

Cross-linking of the starch polymer chains with reagents that could form bonds with more than one hydroxyl group of molecule results in cross-linked starch. Such reactions randomly add inter- and intra-molecular bonds at different locations in the starch granule which helps to strengthen and stabilize the polymers in the granule. Such processes may employ hydrolysis, oxidation, esterification, etherification, phosphorylation or combinations of these methods in a sequential or one-mix procedure to achieve the desired product that meets the required physicochemical characteristic of gelatinization, viscosity, retrogradation, and textural properties for food applications. In some instances, multifunctional reagents capable of forming either ether or ester inter-molecular linkages between hydroxyl groups on starch molecules are used. Reactions usually take place at the primary –OH group of C-6 and secondary –OH of C-2 and C-3 of the glucose units. Epichlorohydrin monosodium phosphate, phosphoryl trichloride, sodium trimetaphosphate, sodium tripolyphosphate, a mixture of adipic and acetic anhydride, and vinyl chloride are the main agents used to cross-link food grade starches. Di-starch phosphate which is a phosphorylated starch is an example of a cross-linked starch. Acetylated distarch adipate, hydroxypropyl distarch phosphate, hydroxypropyl distarch glycerol are other examples of crosslinked starch. The FDA specify that not more than 0.1, 1 and 0.12% DS (w/w of starch) of phosphoryl chloride, sodium trimetaphosphate and adipic-acetic mixed anhydride, respectively, should be used for food grade starch.

Cross-linked starch exhibit increased resistance to processing conditions such as high or low temperatures and pH. Cross-linking reduces granule rupture, loss of viscosity and the formation of a stringy paste during cooking, providing a starch suitable for canned foods and products. Cross-linked starch shows smaller swelling volume, lower solubility and lower transmittance than native starch. While oxidation may increase retrogradation, crosslinking reduces it. Hence a combination of the two chemical modification methods can be used to get the starch with desired balanced characteristics.

Approaches to Modification of Starch

Native starches are modified to improve their physicochemical properties due to different reasons. Different approaches have been reported including physical, chemical, enzymatic and genetic approach. But the most widely used is the chemical approach. For instance, since starch must be gelatinized for it to be digestible in human diet and nutrition, and the process of gelatinizing native starches usually takes appreciable amount of time for granule to swell and form paste of gel as obtained in cooking rice and corn flour porridge, it can be modified to reduce gelatinization time by physical methods such as extrusion, spray-drier and drum dryer, which promote fast starch gelatinization to produce pregelatinized starch. Pre-gelatinized starch exhibit reduced gelatinization temperature and time. The modified starches are usually dries to obtain flours and/or pre-gelatinized starches of long-term stability and quick preparation. Pregelatinized starches are partially or totally soluble in cold water and readily form pastes. It absorbs more water and disperses readily in water than the untreated starch, forming gel at room temperature and less prone to deposit. Using gelatinized starch in food products affects the food qualities and properties, such as, bread volume and crumb ; pastas elasticity and softness, lusciousness and digestibility, tolerance in the properties of beating and cake mixtures, ice creams, doughnuts, growth of sugar crystals in food products ; texture, volume, shelf-live and stability during thawing of cakes and breads. Liquefaction, partial hydrolysis and dextrinization may occur during pregelatinization depending on the processing conditions.

The process of physical modification does not involved any chemical reaction of starch with a modifying reagent and is referred to as physical modification of starch and the products are known as physically modified starches. However, most modifications of starches are performed through chemical processes. The chemical reactions of starch (hydrolysis, esterification, etherification, oxidation and cationization) are generally exploited in the industry to produce converted or modified starches fit for different purposes in the industry.

According to the Food and Nutrition Program (FNP) of the FAO, a modified starch is a food starch which has one or more of its original physicochemical characteristics altered by treatment in accordance with good manufacturing practice by one of the reaction procedures such as hydrolysis, esterification, etherification, oxidation and cross-linking. For starches subjected to heating in the presence of acid or with alkali, the alteration (mainly hydrolysis) is considered a minor fragmentation. Bleaching is also essentially a process resulting in the colour change only. However, oxidation involves the deliberate creation of carboxyl groups. Treatment of starch with substituting reagents such as orthophosphoric acid etc., results in partial substitution in the 2-, 3- or 6-position of the anhydroglucose unit (AGU) unless the 6-position is occupied for branching in amylopectin chain. For cross-linked starch, where polyfunctional substituting agent, such as phosphorus oxychloride, connects two chains, the structure can be represented by Starch—O—R—O—Starch, where R is the cross-linking group and Starch refers to the linear and/or branched structure.

Evolving biotechnological innovations are progressing with enzymatic and genetic modification of starch as a greener alternative to chemical modification due to environmental concerns. Enzymatic modifications basically employ hydrolytic enzymes found in certain bacteria. For instance amylomaltases or α-1,4-α-1,4-glucosyl transferases from Thermus thermophiles and cyclomaltodextrinase (CDase 1–5) from alkalophilic Bacillus sp. While α-1,4-α-1,4-glucosyl transferases breaks existing α-1,4 bonds and make new ones to produce modified starch used in foods and non-foods applications, CDase 1–5 can be used to produce starches which are low in amylose content without changing the amylopectin distribution. The granule of starch-cyclomaltodextrin complex produced special tastes and flavours, as well as light, heat and oxygen-sensitivity stability. Transglucosidase, maltogenic α-amylase and β-amylase have been used to produce resistant starches of various degrees of digestibility. On the other hand, genetic modification employs biotechnology to targets the starch biosynthetic process. Genetic regulation of enzymes such as starch synthetase and branching enzymes, involved in starch synthesis through starch synthase genes are used to produces cereal crops that yield amylose- free starch, high-amylose starch and altered amylopectin structure in starch.

Starch Functionality and its Applications in Food

The reactions of starch are exploited to create different types of modified or converted starched to obtain starches with appropriate physicochemical characteristics such as gelatinization, retrogradation, heat stability, solubility, transmittance, colour, texture, etc., for different industrial applications. The food industry is very mindful of safety of chemical residues hence not all types of modified starched are used in foods. Generally, modified starches are used for adhesion and as binder in battered and breaded foods, formed meat and snack seasonings; as dustings for chewing gum and products produced in the bakery; as crisping cover for fried snacks; fat replacer and juiciness enhancement in ice cream and salad dressings; flavour encapsulating agents in beverage clouds; emulsion stabilizers in beverages, creamers and canned foods; foam stabilizer in marshmallows; gelling agents in gum drops and jelly gum; and as expanders in baked snacks and cereal meals.

Table: Application of chemically modification starches in foods.

Chemical process	Specific treatment	Products	Function	Food application
Hydrolysis	Acid treatment	Acid-hydrolized starch, acid thinned or thin-boiling and fluidity starches.	Reduced hot-paste viscosity, improved gelling or gel strength. Enhanced textural properties.	Gum, pastilles, jellies
	Acid treatment	Dextrinized starch	Increased solubility and gel stability, reduced viscosity and improved emulsification properties. Encapsulate volatiles aromatic compound such as limonene, isoamyl acetate, ethyl hexanoate and β-ionones.	Fat replacer in bakery and dairy products, bakery glazes, protective coating in confectionery. Flavour encapsulator in seasonings.
	NaOH or KOH treatment	Alkaline hydrolysed starch	Increase viscosity	
Oxidation	Sodium hypochlorite oxidized starch	Oxidized starch	Lower viscosity, improved whitening of granules, high paste clarity, low temperature stability, and increased adhesion. Reduces retrogradation of cooked starch pastes.	As binder in battered meat and breading, film former and binder in confectionery, crispy coating in various fried food stuffs, texturizer in dairy products.
Esterification		Monosubstituted starch (starch acetates, starch hydroxypropyl ethers, starch monophosphate esters).	Freeze-thaw stability, improved emulsification properties.	As emulsion stabilizers and for flavor encapsulation in refrigerated and frozen foods.
	Acetylation with acetic acid anhydride	Starch acetate	Increased lipophilicity emulsion stabilizer. Improves quality of any fat/oil-containing products. Reduces rancidity by preventing oxidation. Increase viscosity.	Bulking agent in snack foods, stabilizer and thickener in most foods, improves smoothness and sheen of soups and sauces. Cholesterol-free salad dressings, and flavor encapsulating agents in clouding agents, creamer and beverage. Substitute to gum arabic, egg yolk and caseinates.
	Succinylation with succinic acid anhydride	Starch succinate	Improved viscosity and juice taste. Freeze-thaw stability.	Soups, snacks, and frozen/refrigerated food products. As thickener and in non-gelling custard creams. Meat and fried products to improve juicy or smooth taste and retain flavour.
	Succinylation with OSA	OSA starch	Increased paste viscosity, emulsion stabilizer and lower gelatinization temperature. Reduces glycemic response after consumption of beverages.	Beverage emulsion stabilizers, and mayonnaises. Flavour encapsulating agent for battered meat and meat products.

	Treatment with adipic anhydride	Starch adipate	Higher paste viscosity, clarity and stability.	Thickening agent in foods
	Phosphorylation	Starch phosphate	Better paste clarity, lower gelatinization temperature, higher viscosity, reduced retrogradation, and improved freeze-thaw stability.	Frozen foods
		Distarch phosphate	Stability against high temperature, low pH and shear, and improved firmness of the swollen starch granule as well as improved viscosity and textural characteristic, resistance to syneresis during storage.	As a thickener and stabilizer in foods such as soups and sauces.
Etherification		Etherified starches	Improved clarity of starch paste, greater viscosity, reduced syneresis and freeze-thaw stability.	As stabilizer in wide range of food applications such as gravies, dips, sauces, fruit pie fillings and puddings. Flavour encapsulating agent in beverages clouds.
	Carboxymethylation	CMS	Cold-water solubility	Candy foods, sweets
	Hydroxypropylation	HPS	Improves freeze-thaw stability, water-holding properties, lowers the swelling/pasting temperature, increases paste clarity and reduces gel formation. More stable to prolong high temperatures. Increase solubility.	Salad dressing, ice creams, refrigerated and frozen foods, and dairy products.
	Cationization	Sulfonium starch	Higher dispersibility and solubility with better paste clarity and stability.	
Crosslinking		Crosslinked starches	Higher stability to granules swelling, high temperature, high shear and low pH. Better viscosity and freeze-thaw stability. Volume expander. Delays retrogradation and reduce paste clarity.	As thickener and texturizers in soups, sauces, gravies, bakery and dairy products. Filling in fruit pies and canned foods. In bread and dough products as expander and to improve rheological properties.
		Crosslinked-hydroxypropylated starch	A smooth, viscous, clear thickener and freeze-thaw stability.	Gravies, dips, sauces, fruit fillings and puddings.
Pre-gelatinized starch			Cold-water solubility and thickening.	Instant soups, sauces, dressing, desserts and bakery mixes. Thickener in food that receive minimal heat processing such as pastas.

Baked Products (Bread, Pies, Samosas, Wafers, Biscuits and Sausages)

Baked products like biscuits, pies, bread, cakes wafers and sausages are high density products requiring heat resistant starches. Hence cross-linked starches are used since they are more resistant to oven baking temperatures of 120 ≥ 230 °C. Gelatinized starches are also used in ready-to-eat cereal meals such as corn-flakes, etc. The temperature, humidity and degree of stirring determine the texture and quality of the product.

Confectionery (Candy, Sweets and Sweetmeat)

Oxidized starches have high clarity or transmittance, low viscosity and low temperature stability. It is frequently used in confectioneries for coating candies and sweets since they easily melt.

Gravies, Soups and Sauces (Soups, Sauces, Tomato Paste or Ketchup)

Etherified and cross-linked starches are mostly used. Cross-linked starched have higher stability for granules-swelling, high temperature resistant, high shear stability and acidic conditions stability. They are used as viscosifiers and texturizers in soups, sauces, gravies, bakery and dairy products. Etherified starches have improved clarity of starch paste, greater viscosity, reduced syneresis and freeze-thaw stability. Cross-linked starches are used in wide range of food applications such as gravies, dips, sauces, fruit pie fillings and puddings.

Mayonnaises, Salad Dressing, Ice Cream, Spreads and Beverages

Hydrolyzed and esterified starches are mostly used in salad dressing and beverages. Hydrolyzed starch (acid-modified starches) has lower paste viscosity under cold and hot conditions. Hence they are used in mayonnaises and salad dressing. Esterified starches have lower gelatinization temperature and retrogradation, lower tendency to form gels and higher paste clarity, and are used in refrigerated and frozen foods, as emulsion stabilizers and for encapsulation of beverage clouds. OSA starch is used as emulsifiers in mayonnaises and salad dressings.

Pasta (Spaghettis, Macaroni, others)

Pregelatinized and cross-linked starches are mostly used in pastas. Gelatinized starch affects pastas elasticity and softness, delectableness and digestibility. Crosslinking gives the needed structural firmness to the pasta.

Puddings (Custard, Pap, others)

Pregelatinized starches are used in puddings, instant lactic mixtures and breakfast foods to achieve thickening or water retention without employing heat. They are also used in ready-to-use bread mixtures. They are used where little or no heat is required and the increased absorption and retention of water improves the quality of the product; as an agglutinant in the meat industry; and as a filling for fruit pies.

Thickening and Gelling Agents

Hydrocolloids are frequently used in several foods for thickening and gelling. Hydrocolloids are widely used in many food formulations to improve quality attributes and shelf-life. Hydrocolloids are a heterogeneous group of long chain polymers (polysaccharides and proteins) characterised by their property of forming viscous dispersions and/or gels when dispersed in water. Presence of a large number of hydroxyl (-OH) groups markedly increases their affinity for binding water molecules rendering them hydrophilic compounds. Further, they produce a dispersion, which is intermediate between a true solution and a suspension, and exhibits the properties of a colloid. Considering these two properties, they are aptly termed as 'hydrophilic colloids' or 'hydrocolloids'.

The foremost reason behind the ample use of hydrocolloids in foods is their ability to modify the rheology of food system. This includes two basic properties of food system namely, flow behaviour (viscosity) and mechanical solid property (texture). The modification of texture and/or viscosity of food system help to modify its sensory properties, and hence, hydrocolloids are used as important food additives to perform specific purposes. It is obvious that several hydrocolloids belong to the category of permitted food additive in many countries throughout the world. Various food formulations like soups, gravies, salad dressings, sauces and toppings use hydrocolloids as additives to attain the desired viscosity and mouth feel. They are also used in many food products like ice-cremes, jams, jellies, gelled desserts, cakes and candies, to create the desired texture. Considering their role in the adjustment of viscosity and texture of food formulations, several studies have been conducted in various food systems employing different hydrocolloids either singly or in combination.

Hydrocolloids have a wide array of functional properties in foods. These include thickening, gelling, emulsifying, stabilisation, and controlling the crystal growth of ice and sugar though the basic properties for which hydrocolloids find extensive use are thickening and gelling. Hydrocolloids disperse in water to give a thickening or viscosity producing effect. This water-thickening property is common to all hydrocolloids and is the prime reason for their overall use. The extent of thickening varies with the type and nature of hydrocolloids, with a few giving low viscosities at a fairly high concentration but most of them giving high viscosities at concentration, below 1%.

While all hydrocolloids thicken and impart stickiness to aqueous dispersions, a few biopolymers also have another major property of being able to form gels. Gel formation is the phenomenon involving the association or cross-linking of the polymer chains to form a three dimensional network that traps or immobilises the water within it to form a rigid structure that is resistant to flow. In other words, it becomes viscoelastic exhibiting both characteristics of a liquid and a solid. The textural properties (e.g., elastic or brittle, long or spreadable, chewy or creamy) of a gel vary widely with the type of hydrocolloid used. The other sensory properties such as opacity, mouth feel and taste also depend on the hydrocolloid employed. Hydrocolloids that are commonly used as thickening are starch, xanthan, guar gum, locust bean gum, gum karaya, gum tragacanth, gum Arabic and cellulose derivatives. The gelling type hydrocolloids are alginate, pectin, carrageenan, gelatin, gellan and agar.

Thickening Agents

Hydrocolloids are frequently used in several foods for thickening. The process of thickening involves the non-specific entanglement of conformationally disordered polymer chains; it is essentially a polymer-solvent interaction. Thickening occurs above a critical concentration known as overlap concentration (C*). Below this, the polymer dispersions exhibit Newtonian behaviour but show a non-Newtonian behaviour above this concentration. Hydrocolloids that have been used as thickening agents in various food systems include starch, modified starch, xanthan, galactomannans like guar gum and locust bean gum (LBG), gum Arabic or acacia gum, gum karaya, gum tragacanth and carboxymethyl cellulose (CMC). The thickening effect produced by the hydrocolloids depends on the type of hydrocolloid used, its concentration, the food system in which it is used and also the pH of the food system and temperature.

Table: Important hydrocolloid thickeners used in foods.

Hydrocolloid as a thickener	Properties	Application in food products
Xanthan	Highly shear thinning; maintains viscosity in the presence of electrolytes, high temperature and wide pH ranges.	Soups and gravies, ketcups, instant beverages, desserts, toppings and fillings.
Carboxymethyl cellulose (CMC)	High viscosity but is reduced by adding electrolytes and at low pH.	Salad dressings, gravies, fruit pie fillings, ketchup.
Methyl cellulose (MC) and hydroxypropylmethyl cellulose (HPMC)	Viscosity increases with temperature but independent of pH and electrolytes.	Salad dressings, cake batters, beverages, whipped toppings.
Gum Arabic	Low viscosity gum; shear thinning at low shear rates (<10/sec); near Newtonian behaviour above 100/sec of shear rate.	Fruit juice based beverage, soft drinks.
Galactomannans (Guar gum, Locust bean gum and tara gum)	Very high low-shear viscosity; highly shear thinning; independent of electrolytes but degrade and lose viscosity at high and low pHs and at high temperatures.	Dairy products including ice cream, ketchup, fruit juices, pudding powder, cake batter.
Konjac maanan	Forms highly viscous dispersions which are not influenced by addition of salts; forms thermally irreversible gels with alkali.	Noodles and jelly desserts.
Gum Tragacanth	Swells rapidly in cold or hot water to form highly viscous dispersions, up to 4000 mPas at 1% solids.	Salad dressings, bakery emulsions, fruit beverages, sauces.

Starch is the most commonly used hydrocolloid thickener, the reason being it is relatively cheap, abundant and possibly it does not impart any noticeable taste if used at a low concentration of 2 to 5%. Further, as starch is a common ingredient of many foods we encounter, addition of starch does not offer any foreign taste which may be true for different gums. It is mainly the hydrocolloid providing a base texture in soups and sauces. Thickening of sweet and sour sauces with various polysaccharide combinations like potato starch-xanthan gum and oat starch-xanthan gum has been studied. The evaluation of the thickener performance is considered on the basis of its effect upon sensory properties and rheology. Oat starch-xanthan gum combination has better thickening property compared to potato starch-xanthan gum combination as evaluated from their energy of thixotropy. Both potato starch and xanthan are anionic polysaccharides and hence are thermodynamically incompatible for efficient intermolecular interaction required for thickening. In dessert sauces like strawberry sauce, starch-xanthan combination is found to be suitable as a thickener. It provides stable sensory and textural properties of the sauces for at least three months. While starch thickener used sauces are most stable on storage; potato starch with 0.12% xanthan gum is a superior thickener.

Besides starch and xanthan, other hydrocolloids used for thickening various food systems are summarized in Table. Guar and locust bean gums are galactomannans having different water solubilities due to the difference in the degree of galactose substitution. While guar gum having higher galactose content swells and disperses almost completely in both cold and hot water, locust bean gum (LBG) needs to be heated for complete solubility. Viscosity usually ranges from 6000 to 7500 mPas for guar and 3000–3500 for LBG dispersions at 1% solids. The reason for higher viscosity of guar compared to LBG is the higher molecular weight of guar gum.

Table: Major hydrocolloid gelling agents and their characteristics.

Hydrocolloid as a gelling agent	Characteristics	Application in food
Modified starch	Thermally irreversible opaque gels formed on cooling	Dairy desserts
Agar	Thermoreversible gels on cooling	Bakery products, jellies
κ-Carrageenan and ι- carrageenan	Thermoreversible gels on cooling	Puddings, milk shakes, tofu
Low methoxy pectin	Thermoreversible gels on cooling at acidic pH	Jams, jellies, glazes, milk based desserts
High methoxy pectin	Thermoirreversible gels on cooling at high acidic pH	Jams, jellies
Gellan gum	Thermoreversible highly transparent gels formed on cooling	Water-based fruit flavoured jellies, lemon jelly
Alginate	Thermoirreversible gels do not melt on heating	Restructured foods, cold prepared bakery creams
Methyl and hydroxypropylmethyl cellulose	Thermoreversible gels which melt on heating	Salad dressings, cake batters, beverages, whipped toppings

Carboxymethyl cellulose (CMC), like guar gum, is soluble in either cold or hot water. The concentration, molecular weight and degree of substitution (ds) are important factors for flow behaviour of CMC in aqueous dispersions. Commercial products usually have ds values from 0.7 to 1.5. Dispersions of the gum show shear thinning properties, but products of lower ds are thixotropic and viscosity decreases with increase in temperature. Other cellulosics include methyl cellulose (MC) and hydroxypropylmethyl cellulose (HPMC). Unlike CMC, they form weak gels on heating when the temperature rises above 52 °C for MC and 63–80 °C for HPMC. These are mostly used for binding and shape retention in reformed vegetable products like onion rings, potato croquettes and shaped soya protein. Another hydrocolloid used as a thickener is gum targacanth. This gum swells rapidly, in either cold or hot water to form highly viscous dispersions, up to 4000 mPas at 1% solids, depending on the grade. The dispersions are shear thinning and possess good yield value.

Ketchup is one of the most common food items where the hydrocolloid thickeners are used to influence its viscosity. Sahin and Ozdemir found that addition of LBG, tragacanth gum, guar gum and xanthan gum to ketchup resulted in greater shear thinning properties while CMC showed marginal effect. Consistency index and apparent viscosity increase with the addition of all hydrocolloids, but the increase is highest with the addition of guar and LBG, followed by xanthan and tragacanth and the least with CMC. Koocheki et al. found that though guar gum produced higher apparent viscosity in tomato ketchup as compared to xanthan and CMC, the yield stress is higher for ketchups with xanthan and CMC but lower after addition of guar gum.

The rheological properties and quality of rice cakes formulated with different gums have also been studied. HPMC containing rice cake batters produced lowest apparent viscosity while xanthan and xanthan-guar gum blend containing batters produced highest apparent viscosity. Casson yield stress was lowest for HPMC containing batters. Except HPMC, other gums like xanthan, guar gum and LBG increased the emulsion stability of the cake batter.

Gum arabic (Acacia gum) is widely used in the food industry mainly to impart desirable qualities because of its influence over viscosity, body and texture. Mainly it is used as an emulsifier in beverage emulsions. It is a low viscosity gum and has been found to produce low viscosity at 30% concentration compared to 1% xanthan or CMC at low shear rates. Mothe and Rao in their study on the rheological behaviour of aqueous dispersions of gum arabic have found that dispersions of the gum (4–50%, w/v) exhibit shear-thinning characteristics at low shear rates ($<10\ s^{-1}$) and Newtonian plateaus at shear rate above 100 s^{-1}. The infinite shear rate viscosity (η_α) however, is found to increase with increasing concentration of gum.

Process of Thickening

The application of a hydrocolloid depends on the knowledge and understanding of the process of thickening—an important role of any hydrocolloid. The functional roles of food hydrocolloids are (a) thickening, which involves non-specific entanglement of conformationally disordered chains, and (b) structuring/gelling, which involves specific inter-chain association in conformationally-ordered junction zones.

The question that arises is how hydrocolloids thicken solutions. The viscosity of polysaccharide dispersions arises predominantly from physical entanglement of conformationally disordered 'random coils'. In dilute dispersion, the individual molecules of hydrocolloids can move freely and do not exhibit thickening. In concentrated system, these molecules begin to come into contact with one another; thus, the movement of molecules becomes restricted. The transition from free moving molecules to an entangled network is the process of thickening. Thus, hydrocolloid thickeners can be considered as entanglement of network producers. In cases where no enthalpic polymer-polymer interactions are observed, thickening properties are associated with molecular weight and concentration of the hydrocolloid.

Intrinsic viscosity is an important parameter used to compare the viscosities of dispersions of hydrocolloids. The intrinsic viscosity (η) is directly related to the molecular weight of the hydrocolloid and is given by the well-known Mark-Houwink equation.

$$[\eta] = K.M^{\alpha}$$

Where, M is the molecular weight and K and α are the parameters of Mark-Houwink equation. Hence, higher the molecular weight of a polymer, greater will be the intrinsic viscosity produced by it in aqueous medium. The polysaccharide concentration at which the sharp change in viscosity occurs is referred to as the critical overlap concentration and is denoted by C*. The onset of entanglement for a wide range of neutral and charged polysaccharides is found to occur when $C[\eta] \approx 4$ (i.e., $C^* \approx 4/[\eta]$). In other words, the higher the molecular weight of the polysaccharide, the lower the concentration at which C* is exceeded. Polysaccharide dispersions below C* will typically exhibit near Newtonian steady shear flow and the increase in the viscosity of the dispersion is roughly proportional to the number of molecules present. Above C* entanglement dispersion networks will exhibit shear thinning meaning that apparent viscosity decreases with increasing shear rate. At low shear rates, the rate of disruption of entanglements is slower than or equal to the formation of new entanglements and thus exhibits Newtonian flow. The viscosity during this time is known as the zero-shear viscosity. As shear rate is increased, the rate of disruption also increases and exceeds that of formation. At this point, viscosity begins to decrease sharply as a function of shear rate. Again at very high shear rates, no further disruption can take place and the system once again exhibits Newtonian flow; this is called infinite-shear viscosity. The viscosity of hydrocolloid dispersions is also influenced by the polymer hydrodynamic volume. It increases with molecular mass, chain rigidity and electrostatic charge density, and is greater for linear compared to branched polysaccharides. Thus, amylose will have a higher intrinsic viscosity than amylopectin.

Starch is the most commonly used hydrocolloid thickener, and is used both in the native and modified forms. Cold water does not dissolve starch but as the water is heated, it gradually penetrates the granules causing them to swell up. Thus, a marginal thickening occurs at this stage, and finally near the boiling point, the starch granules burst and release the inner components into the liquid and marked thickening occurs. Other gums are often added to starch to improve the texture and mouth-feel.

Gels and Gelling Agents

Gels may also be defined as a form of matter intermediate between solid and liquid and show mechanical rigidity. They consist of polymer molecules cross-linked to form tangled and interconnected molecular network immersed in a liquid medium, which in food system is water. Food technologists use the word 'gel' for high moisture foods that more or less retain their shape when released from their container. However, the most commonly used definition of gel is a rheological one, obtained from dynamic viscometry. According to this definition, a gel is a viscoelastic system with a 'storage modulus' (G′) larger than the 'loss modulus' (G″). Hydrocolloids form gels by physical association of their polymer chains through hydrogen bonding, hydrophobic association and cation mediated cross-linking and differ from synthetic polymer gels, which normally consist of covalently cross-linked polymer chains. Hence, hydrocolloid gels are often referred to as "physical gels".

Though all hydrocolloids thicken aqueous dispersions, only a comparatively few gums form gels. Also the gels thus formed vary widely in gel character and texture. Hence, knowledge of the conditions required for gelling of particular hydrocolloid dispersion, the characteristics of the gel produced and the texture it confers are very important aspects to design a specific food formulation. The important gums that find application in food as gelling agents include alginate, pectin, carrageenan, gellan, gelatin, agar, modified starch, methyl cellulose and hydroxypropylmethyl cellulose. The gel characteristics of some of the important gelling agents are summarized in Table.

Process of Gelling

The formation of gel involves the association of randomly dispersed polymer segments in dispersion in such a way so as to form a three-dimensional network that contains solvent in the interstices. The associated regions known as 'junction zones' may be formed by two or more polymer chains. The gelation process is essentially the formation of these junction zones. Hydrocolloid gelation can involve a hierarchy of structures, the most common of which is the aggregation of primary inter-chain linkages into "junction zones", which form the basis for the three dimensional network characteristic of a gel. The physical arrangement of these junction zones within the network can be affected by various parameters like temperature, presence of ions and inherent structure of hydrocolloid. For the gelation of hydrocolloids, the three main mechanisms proposed are ionotropic gelation, cold-set gelation and heat-set gelation.

Ionotropic gelation occurs via cross-linking of hydrocolloid chains with ions, typically a cation mediated gelation process of negatively charged polysaccharides. Examples of such systems are alginate, carrageenan and pectin. Ionotropic gelation is carried out by either diffusion setting or internal gelation. In cold-set gelation, hydrocolloid powders are dissolved in warm/boiling water to form a dispersion which on cooling results in enthalpically-stabilized inter-chain helix to form segments of individual chains leading to a three dimensional network. Agar and gelatin form gel by this mechanism. Heat set gels require the application of heat to gel (eg, curdlan, konjac glucomannan, methyl cellulose, starch and globular proteins). It is usually only where heat setting is required in foods (eg, the use of starch in sauces). Heat setting mechanism occurs by unfolding/ expansion of native starch/protein and their subsequent rearrangement into a network.

Role of Junction Zones in Gelling

Junction zones play a very important role in the gelling process of hydrocolloids. They also markedly influence the characteristics and functional behaviour of a particular gel. The number of molecules that form a junction zone is an important gel property determinant. During gelatin, junction zones are formed by three molecules through hydrogen bonding. In κ-carrageenan, six to ten molecules form a junction zone, whereas in ι-carrageenan only two molecules are involved. More the number of molecules in the junction zone, more rigid will be the gel. Hence, the multi-molecule junction zones of κ-carrageenan are more rigid and less easily rebuilt when disturbed by shearing force whereas ι-carrageenan gels have much more flexible textures and are less sensitive to shearing. The number of junction zones and number of molecules in the junction zones and the flexibility of the interrupting segments are important for the characteristics of a set gel. The junction zones of ι-carrageenan and alginate consist of two molecules, but ι-carrageenan gels can sustain more deformation before they break compared to alginate gels of almost the same strength. The thermal behaviour of gels also differs because of the junction zones. Gelatin melts at much lower temperature because the junction zones are only bound by weak hydrogen bonds. On the other hand, it is possible to make alginate gels that do not melt on boiling because of the strength of calcium bridges in the junction zones. One of the major factors influencing the strength of junction zones is their length. Calcium bridging is cooperative, i.e., the bond strength increases more than proportionally with junction zone length. Solvent quality is also another important factor. Hydrogen bonds in high methoxy pectin gels can only be formed if the water activity is sufficiently reduced by addition of sugar.

There are various types of junction zones in polymer gels like stacked double helix junctions in carrageenans, partly stacked triple helices in gelatin and 'egg-box' junction in alginate. In most polymer gels, the junction zones contain a substantial proportion 30% of the polymer material. Various factors affect gel formation by hydrocolloids which include concentration of the gelling agent, pH of the medium, molar mass/degree of polymerization, temperature, ionic composition and solvent quality. Typically used gelling concentrations of different hydrocolloid gelling agents in food are summarized in table.

Table: Concentrations of gelling agent used in different food formulations.

Hydrocolloid	Concentration of gelling agent, % w/w
Agar	1–2
Alginate	1–2
Carrageenan	0.5–3
High methoxy pectin	2–4
Low methoxy pectin	0.1–4
Gelatin	1–5
Gellan	0.5–1.5

In addition to having knowledge of the factors that affect gel formation by hydrocolloids, it is also necessary to characterize the gels formed by them. Microstructural and rheological characterizations are generally done and are often followed by sensory evaluation. Rheological characterization of gels involves characterizing a gel on the basis of various parameters like modulus of elasticity, yield stress, shear modulus, storage and loss modulus, complex viscosity, gel strength and compliance. These parameters are usually determined by conducting tests like compression test, dynamic oscillatory rheometry, creep and texture profile analysis, etc. by employing instruments like universal texture measuring system, controlled stress rheometer. Not only the measurement but also studying the influence of gel forming conditions on the attributes and characteristics forms an indispensable part of the study on gels. These studies, in turn, help in the formulation of food products, wherein hydrocolloids are incorporated as gelling agents. For example, the effect of addition of sucrose and aspartame on the compression resistance of hydrocolloid gels namely, κ-carrageenan gel, gellan gel and κ-carrageenan/locust bean gum gel has been studied. Addition of sucrose results in an increase of true rupture stress in all these gels. However, addition of aspartame at low concentration does not affect the textural compression parameters. In addition, the main factors determining the gel sweetness are related with mechanical properties of gel like gel strength, rupture stress, rupture strain and particularly with the amount of deformation required to break the network and with its resistance to deformation. Besides, co-solutes like sucrose, concentration of hydrocolloid, shear rate and temperature are also important variables that influence the rheological status of hydrocolloid gels.

The non-gelling agents (eg, xanthan and guar gum), and gelling agents (carrageenan and locust bean gum) are commonly combined to achieve increased viscosity or superior properties of food gels, such as higher elasticity. The blending of different polysaccharides offers an alternative route to the development of new textures. The major interest lies in the development of synergistic mixtures with improved or induced gelation.

Both xanthan and galactomannans do not form gels when used singly. But together, they form gels because of synergistic interactions. The mixture of xanthan and galactomannan is one of the oldest and most extensively studied synergistic gelling systems. Xanthan shows quite spectacular synergistic interactions with other non-gelling polysaccharides of galactomannan family leading to increase in viscosity and gel formation. The distinction between increased viscosity and gelation lies on the mannose-galactose ratio of the galactomannan. Xanthan-guar mixtures exhibit synergistic increase in viscosity as do low concentration mixtures with locust bean gum (<0.03%). At higher concentrations, soft and elastic gels are formed with locust bean gum (LBG). Xanthan gum-LBG gel is thermally reversible and it melts between 55 and 60 °C while dispersions of xanthan-guar mixtures can be gelled at room temperature. Mixtures of xanthan and LBG require heating to 90–95 °C. The interaction of xanthan gum with galactomannan is dependent on the ratio of the mixture, pH and ionic environment. Optimum gum ratios are 80:20 for guar gum: xanthan gum and 50:50 for LBG: xanthan gum. Generally, the synergistic interaction with galactomannans is at its maximum in deionised water at neutral pH and it gets reduced at high salt concentrations and low pH.

Xanthan possesses a synergistic interaction with konjac glucomannan (KGM) during gel formation. Their synergism produces thermoreversible physical gels at neutral pH. Best synergism is obtained when konjac to xanthan ratio is 70:30 (w/w). Mixtures of xanthan and KGM are reported to produce 'melt-in-the-mouth' gels having texture similar to that of gelatin. Hence, they provide a useful replacement in applications where 'melt-in-the-mouth' characteristics are important for product quality and where moderate acidity is acceptable or necessary.

Kappa-carrageenan is a gelling agent, the synergistic interactions of which, with other hydrocolloids have been exploited in several food formulations. Hot dispersions of κ-carrageenan-locust bean gum form strong elastic gels with low syneresis when cooled below 50 °C. The maximum interaction, and hence, peak rupture gel strength occurs at ratios between 60:40 and 40:60 of κ-carrageenan to locust bean gum. These polymer combinations are used in large quantities in cooked meats and in gelled pet foods. κ-Carrageenan and clarified locust bean gum mixtures can also be used for glazing of cake or they can be formulated to give clear water dessert gels with an elastic cohesive gel texture like gelatin. A combination of κ- and ι-carrageenan is also able to offer an elastic cohesive gel, which is similar to gelatin in texture. In addition, konjac flour also interacts strongly with κ-carrageenan forming strong elastic gels having rupture strength four times higher than that of κ-carrageenan gel alone.

However, the most important synergistic interactions of κ-carrageenan as a gelling agent is observed in milk gels where it interacts with milk proteins particularly casein. Also interactions between κ-carrageenan, milk proteins and modified starch in sterilised dairy desserts have been studied, wherein modified starch imparts a large influence on the complex modulus of the gel.

Applications of Hydrocolloids as Gelling Agents

The selection of a particular hydrocolloid to be used in a specific food product depends on the characteristics of gelling agent. For example, alginate can form gels without prior heating because sodium alginate is cold water soluble and these cold-formed gels are heat stable. This makes alginate a preferred gelling agent for re-structured foods and for cold-prepared instant bakery custard that are bake-stable. The rapid setting behaviour of alginate gels is also important in restructured foods that are diffusion-set. Alginates as gel forming agents find applications in restructured fruits and vegetables, restructured fish and meat, puddings and desserts, cold prepared bakery creams, fruit preparations and bakery jam. In icings and toppings, fruit pie fillings and table jellies alginates are used but they are incompatible with milk, except in the presence of calcium sequestrants.

Carrageenan is a hydrocolloid that finds maximum application in dairy desserts like puddings, milk shakes, ice cream and chocolate milk because of its ability to form gels in milk at much lower concentrations compared to any other gelling agent. Kappa carageenans form gels in milk at much lower concentrations because of the electrostatic interaction between positively charged region of κ-casein and negatively charged sulphate group of κ-carrageenan that results in increased milk reactivity. Use of carrageenan in tofu (soybean curd) significantly decreases the hardness of tofu when calcium sulphate and calcium acetate are used. At low concentration (2 g/l), carrageenan increases the water holding capacity of soybean protein gels. The other important application of carrageenan is in injected meat, in which only carrageenans can be dispersed in brine without too high viscosity and simultaneous gel formation when the ham has been cooked. Because of the clarity of carrageenan gel and high gelling temperature, it is valued in the preparation of cake glazes and water dessert gels. In addition, the firm and quick setting behaviour of carrageenan gels is made use of in processed cheese systems. Synergistic carrageenan-locust bean gum combinations are being used in table jellies and low sugar, low-acid fruit gels, in which carrageenan has the advantage over pectin that it dissolves well in sugar solutions, whereas pectin has to be pre-dissolved in water. However, under acidic conditions ($pH < 4$), carrageenans require high temperatures for degradation while pectins are acid-stable.

Commercial pectin finds maximum applications in jams and jellies. This is because pectins, especially high methoxyl pectins require >55% sugar solids for gelling. Pectins are the most preferred gelling agents for acidic fruit gels because of being acid stable. To overcome the limitation of high methoxyl pectin, low methoxyl pectins (both commercial and amidated types) are used in the preparation of reduced sugar jams and jellies. Low methoxyl pectins are also used for the production of glazes in bakery industries. Other applications of pectin include desserts, both water gels and milk based gels. Milk and milk products can easily be gelled with low methoxyl pectin because they contain calcium; milk desserts and gelled or thickened milk products like yoghurts can be prepared.

Gelatin exhibits a wide range of functional properties. It can be used as a gelling agent in jellied confectionery. Gelatin gels melt at relatively low temperature (melt-in mouth), and they are slow-setting; all these features make gelatin the preferred gelling agent in yoghurt products, low-fat spreads and sugar confectionery. Marshmallow, an aerated gelled confectionery, uses gelatin as gelling agent for its elasticity and clarity; clarity of gelatin is also the main reason for its use in table jellies. Gelatin has also been used in flavoured gelled milk desserts, either alone or in combination with carrageenan, and also in dessert creams. In dessert creams, it is used to achieve a smooth gel texture.

The food applications of agar are in the areas like bakery products, confectionery, Japanese desserts and confections, meat, fish and poultry products, dairy products, ice cream, peanut butter and beverages. The high melting point of agar gels is improved by the addition of salts. Agar is used in baked goods where it is superior to carrageenans and far superior to gelatin. Agar is also widely used as a gelling agent in jelly confections including traditional Japanese food items and confections like Yokan, Mitsumame, Tokoroten etc. Agar is also used at levels of 0.5–2% to gel canned meat, fish and poultry products.

Gellan gum is unique in the spectrum of gelling agents used in food applications. Rapid setting behaviour, low use level, sparkling clarity of the gel and good flavour release are some of the attributes that make gellan as a preferred gelling agent for food products. It is also used in bakery fillings as partial starch replacement to increase flavour release, in water based dessert gels because of its good clarity and adequate thermal stability. It can also be combined with xanthan or LBG to produce ready-to-eat (RTE) dessert gels. Deacylated gellan gum is used to improve moisture retention, flavour release and storage stability in puddings and to reduce syneresis. The fluid like behaviour of gellan gum at low concentrations may be utilized for the preparation of 'fluid-gels' which are useful in various beverage dispersions. Gellan gum also acts as a very good stabiliser in reconstituted vegetable juice.

Rheology of Thickeners and Gels

Characterisation of gels forms an essential part in the study of thickened dispersions and gels formed by various hydrocolloid gelling agents. Characterisation may involve rheological characterization, structural characterization, microscopic characterization and molecular characterization employing different instruments like viscometers, rheometers, texture measuring systems, differential scanning calorimeters, scanning electron microscope, atomic force microscope, NMR and NIR. Among these, rheological characterisation of sample is most important and generally performed practice since it correlates to the textural attributes of the product, which, in turn, determines its sensory characteristics and consumer acceptability.

In rheological terms, thickened dispersions and gels are viscoelastic materials. They simultaneously exhibit some of the elastic properties of an ideal solid and some of the flow properties of an ideal liquid. Various parameters like apparent viscosity, shear stress, modulus of elasticity, yield stress, shear modulus, storage and loss modulus, complex viscosity and loss angle are used to describe the characteristic rheological behaviour of these food systems.

The rheological behaviour of food products involving hydrocolloids has been studied extensively in both time and frequency domains, using small stress and strain regimes. However, during processing, manufacture and consumption of foods, these systems are subjected to large deformations. Hence, rheological measurement of thickened dispersion/gel has been broadly categorized into small deformation and large deformation testing. These two types of tests give complementary information that does not necessarily correlate. Small deformation testing generally investigates viscoelastic parameters. The most commonly used parameter to describe the rheological behaviour of gels is shear or Young's moduli determined in the so-called linear region (ie, the strain region where the stress required to deform a material is proportional to the strain). These moduli values are independent of the geometry of both the measuring system and the gel sample; consequently these are invaluable for comparing the elasticity of gelling systems. Large deformation tests, on the other hand, are used to measure shear stress, yield stress, apparent viscosity, strain and failure properties of the product. In general, the rheological test methods for the assessment of gel characteristics can be grouped into three types namely, fundamental, imitative and empirical. Each has certain advantages and a few limitations.

Fundamental Methods

Fundamental small deformation tests involve dynamic oscillatory rheometry, creep test and stress relaxation test. In dynamic oscillatory rheometry, the sample is subjected to an oscillatory stress-strain of frequency. The parameters of gel that are measured are storage modulus (G'), loss modulus (G″) and loss/phase angle (δ). Storage modulus (G′) is a measure of the energy that is stored elastically in the gel structure during a periodic application of stress and the loss modulus (G″) is a measure of the energy dissipated or the viscous response. Their ratio tan δ (equal to G″/G′) gives a measure of how much the stress and strain are out of phase with each other. Gels, being viscoelastic in nature, tan δ <1 i.e., δ <90°. These parameters are very important for the rheological characterization of gels. Essentially, solid characteristics are denoted by G′ while G″ indicates liquid characteristics. For weak gel, G′>G″, and thus junction zones can be readily destroyed even at very low shear rate and the network structure is destroyed. For strong gel, G′ >>G″, and both are independent of frequency; lower tan δ values (<0.1) are observed in this case. Another parameter that gives the overall response is the complex modulus (G*). Its value is given by the formula $G^* = \left(G'^2 + G''^2\right)^{1/2}$.

Another important test is the creep test. In a creep experiment, a constant stress is applied and the corresponding strain γt is measured as a function of time. Gel, being a viscoelastic material, responds to the creep test with a nonlinear strain. The important property measured during creep test is the ratio of strain to stress as a function of time, and is referred to as the creep compliance. It describes how compliant a material is; the greater the compliance, the easier it is to deform the material.

Stress relaxation is another important study in the rheology of food. If a gel sample is deformed by a fixed strain and held there over a long time interval, the stress required to maintain this constant strain will gradually decrease due to relaxation of the sample. The sample under testing will only partially recover its original geometry. The relaxation modulus is an important rheological property measured during stress relaxation. It is the ratio of the measured stress to the applied initial strain. Fundamental rheological properties determined at large deformations include fracture, failure and rupture stress/strain. These parameters are usually determined by constant speed experiments such as uniaxial compression and uniaxial tension performed on texture measuring systems. Compression test and tensile tests are generally performed in this study.

Empirical Methods

Compression (or penetration) tests constitute the basis of many small-deformation empirical tests used to measure gel strength. The most common parameter used to measure gel quality is gel strength. The choice of instrument for gel measurement will depend on whether a single-point or multi-parameter analysis. It is governed by factors like high-speed data acquisition, precision and accuracy of results. Large-deformation empirical tests are used to measure another parameter of gels known as rupture strength i.e., the force required to break the sample.

Imitative Methods

The single point measurements, often based on rupture tests, are not representative of the overall mechanical behaviour of gels. A much more comprehensive understanding of gel texture is obtained by analysis of the force-deformation curve generated by compressing a gel sample using a texture measuring system. This instrumental technique is known as texture profile analysis. It is a technique based on compression of free-standing gels twice in succession and is capable of providing both fundamental and empirical data on the mechanical properties of gels. It has the advantage of providing data at both low and high strains allowing gels to be characterised by multiple parameters. These include modulus, hardness, brittleness, adhesiveness, elasticity/springiness, and cohesiveness.

The flowable materials like thin and thick dispersions are conventionally examined employing a viscometer or rheometer. Generally, the shear rate-stress data are collected over a wide range along with the measurement of apparent viscosity, yield stress, zero-shear and high-shear viscosities. A number of available rheological models are used to calculate model parameters like consistency index and flow behaviour index and these parameters help in the characterisation of the samples.

References

- Common-food-emulsifiers: bioenergyconsult.com, Retrieved 27, May 2020
- Gelatin-from-collagen-protein-for-foods-capsules-and-cosmetics, food-industry: delishably.com, Retrieved 13, April 2020
- Alginates-in-food-sodium-alginate-uses-in-food: culinaryphysics.blogspot.com, Retrieved 07, August 2020
- Role-of-pectin-in-food-processing-and-food-packaging, pectins-extraction-purification-characterization-and-applications: intechopen.com , Retrieved 19, March 2020
- What-is-cellulose-1328464: thespruceeats.com, Retrieved 23, Feb 2020
- Chemical-properties-of-starch-and-its-application-in-the-food-industry, chemical-properties-of-starch: intechopen.com, Retrieved 13, Jan 2020

4

Nutrients as Food Additives

The substances that added to food in order to restore the nutrients which are lost during production, for the purpose of correcting dietary deficiencies, are termed as nutritional additives. The concepts elaborated in this chapter will help in gaining a better perspective about the minerals, proteins and vitamins that are added to food as nutritional additives.

The term nutritional additives can be used to mean the addition of vitamins, minerals, amino acids, fatty acids, as well as other pure chemical compounds to food in order to improve or maintain the nutritional quality of foods. However, manufacturers soon discovered that along with an improvement in nutritional qualities, nutritional additives often provide functional qualities.

The earliest use of nutritional additives was to correct dietary deficiencies. In 1833, the French chemist Boussingault recommended the addition of iodine to table salt to prevent goiter. Salt was first iodized in the United States in 1924 when it was shown that the addition of sodium iodine was effective in preventing goiter. Other examples include vitamin D added to milk and vitamin A added to margarine. Nutritional additives can be used to restore nutrients to levels found in the food before storage, packaging, handling, and processing. An early example of this is the enrichment of grain products, corn meal, and rice. Another use of nutritional additives is to improve the nutritional status or correct nutritional inferiority in a food that replaces a more traditional nutritional food, an example would be the fortification of breakfast drink substitutes with folacin and vitamin C. With the advent of nutritional labeling and increased public interest in nutritional properties of food, the food industry rapidly recognized that the addition of nutritional additives can be a selling point. Although it is often thought that the major reason to add nutritional additives to a food supply is to provide nutrients and improve dietary status, nutrients are also added for a variety of other purposes. For example, vitamins C and E may be used for antioxidant properties; beta carotene may be used to provide color. In these cases consumers obtain both a functional and nutrient advantage.

Like all additives, nutritional additives are commercially available in an array of forms such as powders, encapsulated in gelatin, emulsified in oil. The form used depends upon the type of application. Nutrition additives are often protected by protective additives such as antioxidants. Two critical factors in the selection of the form of the additive are stability of the vitamin preparation and its miscibility with the intended food matrix. In order to ensure the second criterion, the point of incorporation of the additive during the food manufacturing process can also be critical.

Minerals as Nutritive Additives

Minerals are inorganic elements that retain their chemical identity when in a food product. Minerals can be divided into two groups, the major minerals and the trace minerals. Major minerals are those present in amounts larger than 5 g in the human body. The seven major minerals include calcium, phosphorus, potassium, sulfur, sodium, chloride, and magnesium. Sulfur, however, is not traditionally used as a nutritional additive. There are more than a dozen trace minerals. Some of the most important trace mineral nutritional additives include iron, zinc, copper, iodine, manganese. Minerals cannot be destroyed by heat, air, acid, or mixing, and only little care is needed to preserve minerals during food preparation. The ash that remains when a food is burned contains the minerals that were in the food originally. Aside from price considerations, the choice of the source depends on three factors:

- Bioavailability of the mineral in a particular salt form.
- Solubility and/or mixability.
- Potential effects on final product properties.

Calcium

Calcium is an essential macromineral, that is necessary for the body in order to fortify bones, teeth and connective tissues, control the contraction and relaxation of muscles, monitor heart activity, normalize blood clotting processes as well as smoothly transmit nerve impulses between cells in the body.

Although four minerals - carbon, hydrogen, oxygen and nitrogen, compose about 96% of the human body by weight, they are not considered essential minerals but are termed as structural minerals. Essential minerals are those that cannot be synthesized by the body and need to be supplied in the form of diet or supplements, which the system requires to perform a specific set of functions.

Calcium comprises about 2% of total body weight in an average healthy adult, is mainly concentrated in the bones, teeth and skeletal assembly, with a small amount stored in the blood and other tissues.

Functions

- Enhances Bone Strength: Calcium in the body is stored in the form of calcium phosphate in the bones in the arms, legs and skeletal muscles of the spinal cord. Right from childhood, adolescence to becoming a young adult and ageing further, calcium is absolutely necessary for maintaining optimal bone health. As bones are the fundamental connective tissues in the body providing structure and stability, encompassing and protecting the more sensitive internal organs such as lungs, stomach, liver, brain, heart and kidney, consuming the mineral calcium in adequate amounts is very important to keep them robust and sturdy.

- Regulates Nerve Signaling: Neurons or nerve cells are the specialised cells in the brain, which are capable of transmitting vital information such as transport of nutrients via blood, response to severe pain, temperature variations, strong tastes and harsh environmental conditions to the muscles, bones and other organs and tissues, so as to evoke the appropriate voluntary action to nurture the body and defend the system from any harm or injury. Calcium in the form of calcium ions forms part of these elaborate biochemical signaling processes, to facilitate the smooth transmission of nerve impulses and maintain overall health and harmony in the body.
- Normalizes Cardiac Muscle Activity: The heart is one of the chief organs in the body. An intrinsic network of cardiac muscles works tirelessly all day, to pump blood in and out of the heart, to various other cells and tissues. The filaments present within the cardiac muscles obtain a charge due to calcium ions, regulating their polarization and depolarization. This in turn aids in proper contraction and relaxation of heart muscles, ensuring the process does not occur too rapidly or slowly. Thus, calcium assists in upholding blood pressure levels within the normal range and augments heart health.
- Promotes Dental Health: Calcium optimizes the resilience of teeth and gums, by working in synergy with two other central essential nutrients, namely vitamin D and phosphorous. This is also the reason why many tooth pastes come infused with calcium salts. Being an inherent hardening element, calcium aids in toughening the jawbones, which defend the teeth and gums against wounds from external physical force and other disease-causing microbes. Calcium is a mandatory nutritional requirement to preserve the good health of teeth and gums for people of all ages.
- Balances Hormone Levels: Calcium circulates in the bloodstream in the form of calcium ions. It plays a key role in standardizing the functions of various hormones in the body, including the reproductive hormones progesterone, estrogen in women and testosterone in men. Calcium is a critical prerequisite for ensuring timely menstrual cycles in young women as well as alleviating symptoms of pain in premenstrual syndrome, and fatigue, weakness in menopause. The intake of calcium gradually increases as women age, with those in the adolescent, pregnant and post-menopausal phases requiring much higher levels of daily intake.
- Food Sources: Calcium is extensively available in a host of natural foods bestowed upon us by Mother Nature. The following foods are incredible sources, for children, young adults, middle-aged and the elderly, to meet their daily requirements of calcium.
    - Dairy: Milk, yoghurt, buttermilk, cheese.
    - Vegetables: Broccoli, cabbage, spinach.
    - Fruits: Orange, kiwi, papaya.

- Nuts and Legumes: Almonds, walnuts, brazil nuts, bean and sprout varieties.
- Fortified Foods: Cereals, juices, breads are incorporated with calcium.

Supplements

In instances where the diet consumed by people on a regular basis is not balanced and food sources of calcium are difficult to obtain, consuming calcium supplements every day vastly improves calcium levels in the body. Calcium in supplements in the form of tablets, capsules and syrups, is usually found in the form of two salts of calcium - calcium carbonate and calcium citrate.

Nevertheless, always seek advice from a professional medical practitioner, to know the exact time and dosage at which calcium supplements must be taken. This averts the risk of any side effects due to the overconsumption of calcium.

Deficiency

When calcium levels in the blood and bones decrease tremendously, it is referred to as hypocalcemia. This negatively affects bone strength, hormonal balance, nerve signal transmission as well as cardiac muscle functioning. Apart from less intake of calcium by means of diet or supplements, lack of adequate vitamin D and phosphorous also influence poor calcium profile in the body.

As calcium is integral to bone structure and function, low calcium levels invariably result in osteoporosis and arthritis, where optimal bone density and flexible joint movement are hampered respectively. Hence, it is crucial to consume sufficient calcium in food or supplemental form, to steer clear of these debilitating deficiency diseases.

Toxicity

Mineral accumulation in the system could give rise to serious repercussions. As calcium is ingested in the system, it is also a necessity to ensure the required amount is utilised to carry out numerous different tasks and the excess calcium is excreted by means of sweat and urine. When calcium levels in the body rise above normal, it results in gas, bloating, constipation and even kidney and heart complications.

It is hence absolutely essential to consult a physician, to find out the current calcium status of the body and consume the essential mineral from food sources and supplements in the appropriate amounts, in moderation.

Phosphorus

Phosphorus is an essential mineral, is naturally present in many foods and available as a dietary supplement. Phosphorus is a component of bones, teeth, DNA, and RNA. In the form of phospholipids, phosphorus is also a component of cell membrane structure and of the body's key energy source, ATP. Many proteins and sugars in the body are phosphorylated. In addition, phosphorus plays key roles in regulation of gene transcription, activation of enzymes, maintenance of normal pH in extracellular fluid, and intracellular energy storage. In humans, phosphorus makes up about

1 to 1.4% of fat-free mass. Of this amount, 85% is in bones and teeth, and the other 15% is distributed throughout the blood and soft tissues.

Many different types of foods contain phosphorus, mainly in the form of phosphates and phosphate esters. However, phosphorus in seeds and unleavened breads is in the form of phytic acid, the storage form of phosphorus. Because human intestines lack the phytase enzyme, much phosphorus in this form is unavailable for absorption. Phosphorus undergoes passive absorption in the small intestine, although some is absorbed by active transport.

Phosphorus and calcium are interrelated because hormones, such as vitamin D and parathyroid hormone (PTH), regulate the metabolism of both minerals. In addition, phosphorus and calcium make up hydroxyapatite, the main structural component in bones and tooth enamel. The combination of high phosphorus intakes with low calcium intakes increases serum PTH levels, but evidence is mixed on whether the increased hormone levels decrease bones mineral density.

The kidneys, bones, and intestines regulate phosphorus homeostasis, which requires maintenance of urinary losses at equivalent levels to net phosphorus absorption and ensuring that equal amounts of phosphorus are deposited and resorbed from bone. Several hormones, including estrogen and adrenaline, also affect phosphorus homeostasis. When kidney function declines, as in chronic kidney failure, the body cannot excrete phosphate efficiently, and serum levels rise.

Although phosphorus status is not typically assessed, phosphate can be measured in both serum and plasma. In adults, normal phosphate concentration in serum or plasma is 2.5 to 4.5 mg/dL (0.81 to 1.45 mmol/L). Hypophosphatemia is defined as serum phosphate concentrations lower than the low end of the normal range, whereas a concentration higher than the high end of the range indicates hyperphosphatemia. However, plasma and serum phosphate levels do not necessarily reflect whole-body phosphorus content.

Sources of Phosphorus

Food

Many different types of foods contain phosphorus, including dairy products, meats and poultry, fish, eggs, nuts, legumes, vegetables, and grains. In the United States, dairy products contribute about 20% of total phosphorus intakes, and bakery products (e.g., breads, tortillas, and sweet bakery products) contribute 10%. Vegetables and chicken contribute 5% each. The absorption rate for the phosphorus naturally contained in food is 40%–70%; phosphorus from animal sources has a higher absorption rate than that from plants. Calcium from foods and supplements can bind to some of the phosphorus in foods and prevent its absorption. According to one analysis, a very high calcium intake of 2,500 mg/day binds 0.61–1.05 g phosphorus. In infants, phosphorus bioavailability ranges from 85%–90% for human milk to approximately 59% for soy-based formulas.

Phosphate additives (e.g., phosphoric acid, sodium phosphate, and sodium polyphosphate) are present in many foods, especially processed food products. These additives are used for such purposes as preserving moisture or color and enhancing and stabilizing frozen foods. Foods containing these additives have an average of 67 mg more phosphorus per serving than similar foods not containing the additives, and these additives contribute to overall phosphorus intakes in the United States.

Phosphate additives are estimated to contribute 300 to 1,000 mg to total daily phosphorus intakes, or about 10%–50% of phosphorus intakes in Western countries. The use of phosphate additives is rising, as are the amounts of these additives in foods. The absorption rate for the phosphorus in phosphate additives is approximately.

Magnesium

Magnesium (Mg^{2+}) is an essential nutrient that is involved in many key metabolic reactions such as energy production, glycolysis, and the synthesis of nucleic acids and proteins. It is also important for oxidative, immune, and neuromuscular functions and bone development. Magnesium assists in maintaining electrolyte balance and calcium, sodium, and potassium homeostasis, all of which are essential for stabilizing excitable membranes. It acts as a calcium channel antagonist, stimulates the production of vasodilator prostacyclins and nitric oxide, and alters vascular responses to vasoactive agonists. The adult human body contains ~25 g of magnesium, ~50–60% of which resides in the bones; most of what remains is present in soft tissues, and <1% resides outside of cells.

Body magnesium is regulated physiologically through 3 main mechanisms: Absorption through the gut, renal excretion after filtration and reabsorption, and exchange from the large pool of bone magnesium. Magnesium is relatively well absorbed by the gut; oral bioavailability varies from 35% to 70% and depends on a variety of factors such as the form of the magnesium salt (organic compared with inorganic), its rate and extent of uptake from the intestine into the blood, and its transfer into tissues because magnesium is primarily an intracellular cation. The absorption rate increases when dietary intake is low.

On a daily basis, ~2.4 g of magnesium is filtered by the kidney at a rate proportional to the plasma concentration, and any excess is rapidly excreted. Adequate kidney function generally prevents toxic systemic and circulatory levels. Urinary magnesium is a relatively good indicator of magnesium intake, and urinary excretion <80 mg/d indicates a risk for magnesium deficiency.

Deficiency

Magnesium deficiency, as measured by serum magnesium levels, due to low dietary intake in otherwise healthy people is uncommon. Although age can negatively affect systemic magnesium levels as absorption from the gut decreases and renal magnesium excretion increases. Hypomagnesemia, typically defined as a serum magnesium concentration <0.75 mmol/L, may result from a number of conditions, including chronic inadequate intake of magnesium, chronic diarrhea, malabsorption, chronic stress, alcoholism, and the use of medications such as diuretics, antacids, proton pump inhibitors, or aminoglycoside antibiotics. The most common symptoms of hypomagnesemia are somewhat nonspecific and include muscle weakness, muscle cramps, and increased irritability of the nervous system with tremors or muscle spasms. Symptoms are usually mild or not present when hypomagnesemia is between 0.5 and 0.7 mmol/L but become more apparent and/or severe when serum magnesium falls below 0.5 mmol/L.

Although overt signs of clinical magnesium deficiency have not been routinely recognized in the healthy population, relatively low magnesium intake and/or magnesium status have been associated with chronic health issues, including cardiovascular disease, type 2 diabetes, osteoporosis, pulmonary disease, depression, migraine headaches, inflammation, and tumor development (2,6). In most cases, these observations did not correlate with a deficiency in serum magnesium, raising the question of the prevalence of "subclinical" magnesium deficiencies. Individuals within the normal serum magnesium range can be low in tissue magnesium, and others with adequate tissue magnesium can be low in serum magnesium. Thus, the concept of chronic latent magnesium deficit has arisen to describe those with low tissue magnesium levels who show serum magnesium in the normal range.

Approximately half of the US population has been shown to consume less than the daily requirement of magnesium from foods. The 2015 Dietary Guidelines Advisory Committee considered magnesium to be a shortfall nutrient that was underconsumed relative to the estimated average requirement (EAR) for many Americans. In the 2007–2010 NHANES, the percentage of teens aged 14–18 y who consumed less magnesium than their EAR was 75% for boys and 87% for girls. Among elderly persons aged ≥71 y, these values were 79% for men and 70% for women.

Food Sources

Magnesium is present in fruits, vegetables, whole grains, legumes, nuts, milk, meat, fish, and in fortified foods such as breakfast cereals. Magnesium is also present in tap, mineral, and bottled waters at varying concentrations; however, most deionized bottled waters sold in the United States contain zero magnesium. Dairy foods have been identified as a key source of magnesium, contributing 17% of the magnesium in the diet for individuals aged >2 y. Nuts, seeds, and soy foods are also good dietary sources of magnesium. The 2011–2012 NHANES reported that 19% of persons aged >2 y consumed supplements with magnesium that varied from 5% to 37% in the various age and gender groups. In the adult population of dietary supplement users, 6% (4% of men and 7% of women) exceeded the tolerable upper intake level of 350 mg/d for magnesium. Mild gastrointestinal effects may occur in a small percentage of individuals at intake levels of 360–380 mg/d, although most individuals do not show such effects even when receiving substantially more than the upper intake level of 350 mg/d.

Potassium

Potassium is a vital trace mineral required by the body to perform numerous important functions such as relaying nerve signals and regulating muscle contractions. In fact, it is the third most abundant mineral found in the human body, after calcium and phosphorous. Potassium reserves are primarily present in cells, of which at least 80% of the mineral is found in muscle cells, while the remaining 20% is stored in bone tissue, liver as well as red blood cells.

The method by which potassium facilitates various bodily functions is by dissolving when in water, to act as an electrolyte. An electrolyte is a substance that splits up into either positive or negative ions in water, to conduct electrical impulses. Potassium breaks down into positive ions in water to enable the smooth passage of nerve signals between organs and tissues in the body.

Potassium is a key mineral required by the body for cellular functions, due to its significant tasks in controlling the fluid volumes within a cell. Along with sodium, another critical electrolyte that regulates the flow of liquids outside cell membranes, potassium helps in preserving the transport of nutrients across the cellular borders.

Functions

Potassium is responsible for many crucial roles in the body, including:

- Maintaining normal blood pressure.
- Transmitting nerve signals between organs.
- Controlling muscle contractions.
- Ensuring optimal water balance within the system.
- Balancing pH in the body between acidity and alkalinity.
- Upholding accurate heart rate i.e. pulse.
- Regulating proper digestion processes.
- Preventing stroke and heart disease.
- Sustaining regular heart muscle activity.

Food Sources

Although potassium is required only in minor quantities as part of the daily diet for children and adults, it is still a significant component of nutrition. Wholesome and balanced meals that consist of ample sources of potassium guarantee the prevention of heart ailments and fluctuations in blood pressure.

Natural food sources of potassium are abundant. The foods which are rich in potassium content comprise:

- Vegetables like green leafy cruciferous plants – spinach, kale broccoli, as well as carrots, potatoes, cucumbers, sweet potatoes and beetroots.
- Fruits such as bananas, apricots, oranges and grapefruit.
- Whole grains like brown rice, wheat bread and oat bran.
- Seeds and nuts including hazelnuts, pistachios, sunflower seeds, flax seeds and almonds.

Deficiency

The normal body levels of potassium lie in the range of 3.5 – 5 mmol per litre of blood serum.

When potassium levels drop significantly below this range i.e. lower than 2.5 mmol/litre, a deficiency disorder arises, termed as hypokalemia.

A person with hypokalemia displays prominent external indications including exhaustion, extreme fatigue even after resting, soreness in muscles and problems with digestion, like bloating, flatulence and constipation.

As the quantities of blood potassium fall dangerously low, much below 2.5mmol per litre, then the severity of symptoms is much higher. The affected individual exhibits partial paralysis in certain muscles and even breathing difficulties and must be given prompt medical care. It is hence advised to go for regular health checks, at least once every 6 months, to obtain current blood test results and ensure that bodily potassium levels are at optimal range.

Toxicity

When the potassium quantities in the system rise to alarmingly high levels, between 5.1 – 6 mmol per litre, it is considered to be excessive, leading to hyperkalemia. In instances wherein the potassium amounts in blood serum surpass 6mmol per litre, it is a very risky situation, especially for those with chronic kidney illnesses.

Hyperkalemia usually occurs only when potassium supplements are consumed in excess and rarely due to too much intake from dietary sources. Nevertheless, it can cause heart complications, chest pain and respiratory distress and must be treated by a doctor at once.

The best way to keep potassium levels in check is to go for routine medical screenings and incorporate healthy changes to diet and avoid supplements, to restore normal body concentrations of the mineral.

Sodium

Sodium is a mineral that your body must have in order to function properly. The primary source of dietary sodium is sodium chloride, or salt, more than three-quarters of which comes from processed foods. Although sodium is vital to a number of routine body functions, too much can have adverse effects, particularly for people who are sensitive to sodium. Excessive sodium can cause hypertension, which in turn can lead to other health problems.

Nerve and Muscle Function

Sodium is a mineral that carries an electrical charge, known as an electrolyte. Electrolytes facilitate muscle contraction and nerve cell transmission. Ions of sodium, potassium and chloride trigger muscle contractions and nerve impulses when they shift places across cell membranes. A nerve cell at rest has positively charged potassium ions inside the cell and is surrounded outside the cell by positively charged sodium ions and negatively charged chloride ions. When stimulated, potassium ions rush out of the cell as sodium ions rush in, creating an electrical signal or nerve impulse. A similar scenario occurs during the contraction of muscles.

Fluid Balance

Sodium also works in concert with potassium to maintain normal water balance in the body. Each of the minerals chemically attracts water to itself, thus assuring that optimal levels of hydration are maintained both inside human cells and outside the cells, in the extracellular spaces that surround them. In healthy people, the body has a built-in mechanism to guard against the effects of occasional excess levels of sodium, but continued intake of high amounts of sodium can eventually override this safety valve and lead to hypertension.

Blood Volume

Closely related to sodium's role in the maintenance of normal fluid levels is the part it plays in controlling your body's blood volume and thus blood pressure. Your body constantly monitors sodium concentrations and blood volume, according to the Merck Manual Home Health Handbook. If either blood volume or sodium levels get too high, your body stimulates your kidneys to excrete excess sodium, returning blood volume to normal levels.

Chloride

Chloride is found in many chemicals and other substances in the body. It is one of the components of salt used in cooking and in some foods.

Function

Chloride is needed to keep the proper balance of body fluids. It is an essential part of digestive (stomach) juices.

Food Sources

Chloride is found in table salt or sea salt as sodium chloride. It is also found in many vegetables. Foods with higher amounts of chloride include seaweed, rye, tomatoes, lettuce, celery, and olives. Chloride, combined with potassium, is also found in many foods. It is most often the main ingredient in salt substitutes. Most Americans probably get more chloride than they need from table salt and the salt in prepared foods.

Side Effects

Too little chloride in the body can occur when your body loses a lot of fluids. This may be due to heavy sweating, vomiting, or diarrhea. Medicines such as diuretics can also cause low chloride levels. Too much sodium-chloride from salted foods can:

- Increase your blood pressure.
- Cause a build-up of fluid in people with congestive heart failure, cirrhosis, or kidney disease.

Iron

Iron performs many important functions in the body. It is primarily involved in the transfer of oxygen from the lungs to tissues. However, iron also plays a role in metabolism as a component of some proteins and enzymes.

Iron is toxic to the body in its free state. It is associated with proteins either through ligand binding or by being incorporated into a porphyrin group - a ring-shaped molecule. A complex of the ferrous form of iron and protoporphyrin IX is known as heme. Heme iron is found in proteins connected with oxygen transport, including hemoglobin and myoglobin. Non-heme iron can be found in proteins connected with oxidative phosphorylation and in iron storage proteins like transferrin and ferritin.

Products containing iron. Healthy eating.

Hemoglobin and Myoglobin

About 70 percent of iron in the body is found in hemoglobin and myoglobin. Hemoglobin is the protein in red blood cells responsible for carrying oxygen to the tissues from the lungs. Myoglobin is a protein found in muscles that is used for storage of oxygen.

Hemoglobin is the oxygen transport system found in the red blood cells of all vertebrates and some invertebrates. In humans, hemoglobin is made up of four globular protein subunits. The four subunits form a pocket that binds a heme group.

Oxygen binds to the iron atom within the hemoglobin molecule in the lungs to form oxyhemoglobin. This occurs in the capillaries of the lung alveoli. It is released at its destination in the cells. Hemoglobin carries CO_2 back to the lungs to be exhaled as waste, but CO_2 binds to the protein portion of the hemoglobin molecule, not to the bound iron in the heme group.

Like hemoglobin, myoglobin binds iron within a heme group. However, structurally, it is much simpler, consisting of a single polypeptide chain of 154 amino acids. It is found only in cardiac myocytes and oxidative skeletal muscle. Myoglobin is an oxygen storage protein. In marine mammals, it provides an oxygen supply for extended periods when the animal is diving under water. At those times, myoglobin releases oxygen to sustain aerobic metabolism in the muscle. In humans, myoglobin levels have been shown to be increased at high altitudes.

Iron-dependent Enzymes

A large number of enzymes require iron as a cofactor for their functions. Among the most significant of these are enzymes involved in oxidative phosphorylation, the metabolic pathway that converts nutrients to energy. The cytochrome enzymes bind heme iron, and some protein complexes in the oxidative phosphorylation process have iron-sulfur centers that are crucial to their function.

Ferritin and Transferrin

Dietary iron is stored within a protein complex called ferritin. Ferritin has 24 subunits that form a capsule around the bound iron atoms. Each complex binds 2000 to 45000 iron atoms. Another protein, transferrin, made in the liver, transports iron within the blood to other locations for storage. The main locations of iron storage in the body include the liver, skeletal muscle, and reticuloendothelial cells. If the storage capacity of these cells is exceeded, iron is deposited near the ferritin-iron complexes in the cells. These deposits are called hemosiderin. Iron in hemosiderin is not available to the cell. Hemosiderin deposits may be found in the body following a hemorrhage.

The overall utilization of iron in the body is regulated by ferritin and transferrin mRNAs that contain iron responsive elements (IREs). Iron homeostasis requires ascorbic acid (vitamin C), which stimulates the absorption of dietary iron, and assists with uptake of transferrin-bound iron in the plasma. Ascorbic acid also stimulates synthesis of ferritin, while inhibiting ferritin degradation and flow of iron out of the cell.

Zinc

Zinc is a vital trace mineral required by the human body to perform numerous important functions such as maintaining defence mechanism and regulating DNA synthesis. It was first discovered as a metallic element in the 1800s, but in the next century, the fact that it has a central role in human physiology as well was unravelled, by many researchers and scientists in the West.

Zinc, although needed only in very minute amounts by the human body, is very essential for fundamental metabolic activities as well as ensuring normal growth and development. Moreover, zinc is the second most abundant trace mineral in the body after iron, being present in practically every single cell in the system. The World Health Organisation enlists zinc as part of the micronutrient mixture in its List Of Essential Medicines, due to the indispensable functions mediated by zinc in the human body.

Functions

Zinc is necessary for carrying out many primary functions in the body such as:

- Bolstering immunity.

- Enhancing nervous system function, brain activity, memory and concentration.
- Lowering inflammation in certain conditions in the body like acne.
- Accelerating wound healing process, in instances of tissue injury.
- Preventing chronic diseases in old age, like diabetes, heart disease and age-related macular degeneration, by fostering healthy ageing process.
- Preserving a normal sense of taste and smell.
- Promoting optimal cell growth, division as well as DNA and protein synthesis.

Food Sources

Since zinc cannot be synthesized in the body, it must be obtained as part of the diet. And nature has bestowed upon us countless foods that are rich in zinc. Some of the food sources that offer profuse amounts of the trace mineral zinc include:

- Legumes such as channa, kidney beans, chickpeas.
- Nuts and seeds like almonds, cashews, pumpkin seeds.
- Dairy products including milk, cheese and yoghurt.
- Whole grains like quinoa, oats.
- Vegetables including mushrooms, peas and greens like spinach, kale.

Deficiency

A deficiency in the quantity of zinc being consumed as part of diet causes many symptoms of discomfort and hampers the functioning of the immune system. Zinc deficiency invariably results in sudden weight loss, opens sores and wounds that do not heal, decreased appetite, diarrhoea and lowered sense of smell and taste.

Toxicity

Excessive intake of zinc through food can result in dangerous levels of the metallic substance accumulating in the body. This causes grave complications such as abdominal cramps, high blood cholesterol levels and severe headaches that affect normal brain functions. Hence, it is advised to consume only the recommended daily intake of zinc and not ingest surplus amounts, in order to steer clear of its toxic effects.

Iodine

An essential mineral, iodine is used by the thyroid gland to make thyroid hormones that control many functions in the body including growth and development. Because your body does not produce iodine, it needs to be supplied in the diet. When iodine intake is poor, the body cannot produce enough thyroid hormones.

Iodine deficiency in pregnancy is a worldwide problem and has become a global public health concern since it is identified as the leading cause of preventable brain damage in newborns and infants due to inadequate intake by mothers and infants. Major international efforts are being made to help reduce the problem, mainly through the use of iodized salt and supplements.

Until the early 1900s, iodine deficiency was a common problem in the United States but was significantly improved with the addition of iodine to table salt. Hypothyroidism, thyroid gland enlargement (goiter) and weight gain are other conditions that may result from too little iodine in the diet. Many pregnant women in the U.S. continue to have insufficient iodine intakes, especially those who have low intakes of dairy, seafood and iodized salt.

Iodine and the Brain

With iodine deficiency listed as the leading cause of mental retardation around the world, iodine is an important component for healthy brain development. The most damaging consequences are on fetal and infant development of the brain when deficiency can cause irreversible brain damage that lasts a lifetime. Brain damage, cretinism, mental retardation and other conditions are additional risks.

It is a crucial nutrient throughout life, but especially during pregnancy, infancy and childhood when thyroid hormones regulate growth in the developing brain. Less severe iodine deficiency manifests as below average IQ in children, including impaired brain function in adults too. During childhood, iodine deficiency is often associated with goiter and also linked to reduced intellectual and motor performance, as well as an increased risk for ADHD in children.

Iodine Requirements

A teaspoon of iodine is all a person requires in a lifetime, but because iodine cannot be stored for long periods, small amounts are needed regularly. The Institute of Medicine, or IOM, recommended dietary allowances for iodine are:

- 1 to 8 years old: 90 micrograms.
- 9 to 13 years old: 120 micrograms.
- 14 years and older: 150 micrograms.
- Pregnant: 220 micrograms.
- Lactating: 290 micrograms.

Best Sources of Iodine

Iodine fortification is what most countries rely on to encourage adequate dietary intake. In the more than 70 countries that iodize salt, it generally serves as the major source of iodine intake. One-fourth of a teaspoon of iodized salts has about 100 micrograms of iodine. Note that the salt used in processed foods, which is the major source of salt for most Americans, typically does not contain iodine. If salt used in a processed food contains iodine, it will be listed in the ingredients list of that food. Focus on decreasing the amount of salt consumed from processed foods and get your sodium from iodized salt.

Seaweed, salt water fish and seafood are natural sources of dietary iodine. Dairy products also supply iodine in the diet at varying levels. During lactation, the breast concentrates iodine in milk, so breast milk tends be a good source of iodine as long as the mother's iodine intake is adequate.

Plants grown in iodine-rich soil are also good sources; however, this is not a reliable source of iodine since there is no way of knowing whether produce purchased in grocery stores is grown in iodine-rich soil.

Iodized salt usually adds less than about 300 micrograms iodine daily to the diet. Most multivitamin mineral supplements contain 150 micrograms of iodine. With the safe upper limit of daily iodine intake for adults set at 1,100 micrograms by the IOM, it is unlikely to hit an excess amount when including a multivitamin and including natural sources of dietary iodine.

The trend of eating less table salt, dairy and bread has some experts concerned that iodine deficiency could be on the rise again. Eating a healthy, balanced diet that includes iodine-rich foods and iodized salt is key to good health. Prenatal vitamins containing iodine can help meet nutritional needs for pregnant and lactating mothers.

Manganese

Manganese is an essential dietary nutrient, being required by the body only in minute amounts and thus classified as a trace mineral. Although it is needed in very less concentrations as part of the daily food regime, it plays an important role in many biochemical reactions taking place within the system and coupled with calcium, vitamin D, reinforces bone strength.

It is a key component of several enzymes in the body, besides actively participating in the metabolism of carbohydrates, amino acids and fats, including cholesterol. Along with vitamin K, manganese regulates blood clotting mechanisms for efficient wound healing processes.

Manganese is mainly obtained from various foods, while only at times being procured from supplements and once ingested, is assimilated by the small intestine. While some mineral content flows freely as manganese ions, most of it gets bound to systemic transport proteins and distributed to other organs in the body.

Around 25 to 40 per cent of manganese is stored in the bones, while the remainder is concentrated in the kidneys, liver, pancreas and brain. After being utilised for bodily functions, manganese is primarily eliminated via bile into fecal matter, with a very small amount being passed from urine. Only a minor proportion of manganese is reabsorbed after being utilised by cells and tissues, for carrying out numerous vital functions.

Functions

Manganese performs several crucial functions in the human body. Being attached to proteins as manganese ions or in the free form, it is involved in the following chief operations:

- Fortifying bones and contributing to bone mineral density, along with calcium, vitamin D.
- An element in the enzyme SOD – Superoxide Dismutase, thereby slowing down ageing and decline in body strength.
- Possesses antioxidant properties to avert heart disease.
- Balances levels of good HDL cholesterol and bad LDL cholesterol in the body, aside from effectively processing triglycerides.
- Lowers the risk of acquiring chronic conditions like type 2 diabetes mellitus and some forms of cancer.
- Decreases inflammation in the system, combined with glucosamine and chondroitin, to alleviate joint pain in osteoarthritis, rheumatoid arthritis.
- Holds vasodilator properties, to efficiently transport blood to the brain, enhance nervous system functioning and avoid epilepsy, seizures.
- Accelerates wound healing mechanisms, working in synergy with vitamin K, thereby assisting in blood clotting and arresting excessive blood loss.
- Preventing osteoporosis and decline in spine structure in older, post-menopausal women, in conjunction with zinc, copper.

Food Sources

There is currently no minimum value ascertained for daily requirement of manganese in the diet. For a normal, healthy individual, a food intake not exceeding 10 mg of manganese per day is considered to be ideal and safe, for at any given point, the body holds reserves of 10 – 20 mg of manganese. A plethora of foods supply manganese in abundant amounts, to effectively meet the average needs of the body for this trace mineral. These include:

- Fruits like pineapple, acai, avocado, coconut, blueberries.
- Green leafy cruciferous veggies of spinach, kale and starchy root vegetables like sweet potato.
- Whole grains and cereal such as wheat, barley, oats, quinoa, rice.
- Legumes including pinto beans, kidney beans, soybeans.
- Nuts like almonds, pecans, walnuts.

Deficiency

In most healthy individuals, manganese deficiency is a very rare occurrence. Despite this, if an individual reports certain symptoms and upon assessing their mineral status, a lack of manganese is identified, then supplements are given. These can be in the form of oral medications or administered intravenously, in case bodily absorption is obstructed due to certain health anomalies.

In children, an insufficient amount of manganese in the system triggers bone malformation, weakness, inadequate growth and development. A less than normal level of manganese in men prompts situations of rashes, redness in skin, along with loss of pigments in hair and reduced serum cholesterol quantities. In women, a manganese deficiency displays prominent mood swings like in the case of cyclothymia, besides complications of premenstrual syndrome.

Toxicity

Manganese toxicity, in most instances, does not occur owing to dietary intake from food sources or even ingesting supplements. The reports of excessive manganese mostly arise due to environmental exposure to the mineral dust, like with people in mining, smelting, welding professions. A condition known as manganism, inhaling manganese fumes is very dangerous, for the body directly delivers the mineral to the brain without properly absorbing the vapor substances.

The indications of manganism are rather similar to that of Parkinson's disease, negatively influencing the brain cells and nerve impulse transmissions within the body. This invariably leads to muscular spasms, breathing distress and lung damage, as well as delusions in the mind, sudden shivers in the spine. Although very rare, comparable symptoms develop in people who have consumed surplus amounts of manganese by means of diet, supplements too and it is thus advised to seek professional medical recommendations, prior to taking supplements.

Moreover, immediately consulting with a doctor as soon as any signs of excessive manganese in the body or manganism are noticed is crucial, to provide timely treatment of the affected person and assure their complete recovery.

Vitamins as Nutritional Additives

Vitamins are organic compounds that facilitate a variety of biological processes. Since the body cannot manufacture vitamins they must be obtained naturally from foods or added to foods. Vitamins are classified as water soluble or fat soluble depending on their characteristics. Water soluble vitamins show poor solubility in water; they are not stored in body tissues in high amounts and excesses are generally excreted in urine. Fat soluble vitamins are linked with lipid metabolism and are generally insoluble or poorly soluble in water. Since fat soluble vitamins can be stored in the human body it is less important to have a daily supply of these vitamins. Excess amounts of fat soluble vitamins can in some cases lead to toxic effects. The fat soluble vitamins are vitamins A, D, E, and K. Vitamins A, C, and E are also classed as antioxidants because they have the capacity to protect the body from free radicals.

Vitamin A

The active principal component of liver oil is vitamin A. Adequate intake of vitamin A is essential for normal vision, growth, cellular differentiation, reproduction, and integrity of the immune system. Deficiency of vitamin A is commonly seen as night blindness and/or dry and lustreless corneas. Vitamin A is one of the major public health problems in less industrialized countries. In fact, it is the major cause of blindness in the developing world. Inadequate amounts are also associated with protein-calorie malnutrition, low intake of fat, lipid malabsorption syndromes, and febrile diseases.

Toxicity: In amounts several times higher than the recommended dietary allowance, vitamin A will cause Toxicity in humans. With children, signs and symptoms of acute vitamin A Toxicity include anorexia, bulging fontanelles, drowsiness, increased intracranial pressure, irritability, and vomiting.

Occurrence and properties: Preformed vitamin A is found in foods of animal origin, either in storage areas such as the liver or associated with the fat of milk and eggs.

In liquid form, vitamin A is a light yellow to red oil that may solidify on refrigeration. In this form it is very soluble in chloroform and in ether. It is soluble in absolute alcohol and in vegetable oils, but is insoluble in glycerin and in water. In solid form, it may have the appearance of the diluent that has been added to it. It may be nearly odorless or have a mild fishy odor, but it has no rancid odor or taste. In solid form it may be dispersible in water. It is unstable to air and light. Vitamin A should be stored in a cool place in tight containers, preferably under an atmosphere of an inert gas, protected from light.

Commercial forms: Vitamin A is available in pure form by chemical synthesis as vitamin A palmitate or the acetate, or recovered from molecularly recovered fish oil.

Carotenoids

Although not vitamins per se, carotenoids serve as a precursor to provide vitamin A. This provitamin is converted by the body to vitamin A with various degrees of efficiency. The yellow, red, orange, and violet pigments of carotenoids are largely responsible for the color of many vegetables and fruits. Carotenoids function not only as color and nutrient compounds but also as antioxidants. Although carotenoids may have favorable effects and may reduce the risk of some diseases, since carotenoids are not essential nutrients there is no need to use the term "carotenoid deficiency".

Toxicity: A yellow, jaundice like coloration of the skin is evident in individuals who routinely ingest excessively large amounts of carotenoids.

Occurrence and properties: Provitamin A carotenoids are synthesized exclusively by higher plants and photosynthetic microorganisms. Carotenoids are mainly obtained from plant sources such as carrots, green leafy vegetables, spinach, oranges, and tomatoes. Animal sources include calf liver, whole milk, butter, cheddar cheese, and eggs.

Carotenoids are soluble in most organic solvents but not in water, acids, or alkalis. They are sensitive to oxidation, isomerization, and polymerization when dissolved in dilute solution under light and in the presence of oxygen.

Commercial forms: Beta carotene is available as red crystals or crystalline powder. Carotenoids are highly sensitive to loss by oxidation and should be protected with appropriate packing materials and storage conditions.

Vitamin D

The primary function of vitamin D is to maintain serum calcium and phosphorus concentrations in a range that supports cellular processes, neuromuscular function, and bone ossification. Vitamin D was found to prevent and cure rickets, a disease associated with malformation of bones. Victims of rickets have traditionally been poor children in industrialized cities where exposure to sunlight has been limited.

Toxicity: Excessive amounts of vitamin D are not normal unless there is excessive use of supplemental vitamins. Toxicity includes hypercalcemia, hypercalciuria, anorexia, nausea, vomiting, thirst, polyuria, muscular weakness, joint pains, diffuse demineralization of bones, and general disorientation.

Occurrence and properties: Food sources are mainly animal products in unfortified foods. Saltwater fish such as herring, salmon, and sardines and fish liver oils are good sources. Other major sources of fortified foods include butter, margarine, cereals, and chocolate mixes. Vitamin D can be produced in the skin after exposure to sunlight. However, in areas with limited sunlight, air pollution, and large cities where buildings block adequate sunlight, the body's ability to synthesize sufficient amounts of vitamin D may be compromised.

Vitamin D is a stable vitamin but is affected by air and by light. Both vitamin D2 and vitamin D3 should be stored in hermetically sealed containers under nitrogen in a cool place and protected from light. Trace metals such as copper and iron will act as pro-oxidants.

Commercial forms: Both vitamin D2 and vitamin D3 are white, odourless crystals. Commercially available forms include fat soluble crystals for use in high fat content foods and encapsulated, stabilized versions of the fortificant suitable for use in dry products to be reconstituted with water.

Vitamin E

Vitamin E is the major lipid-soluble, membrane-localized antioxidant in humans. First discovered in 1922, vitamin E acts in foods to prevent the peroxidation of polyunsaturated fatty acids. In the gut, it enhances the activity of vitamin A by preventing oxidation in the intestinal tract. Vitamin E, at the cellular level, also appears to protect cellular and subcellular membrances from deterioration by scavenging free radicals that contain oxygen.

Sources of dietary vitamin E are widely available and as a result deficiencies are relatively uncommon. Deficiencies that do occur are usually associated with malabsorption or lipid transport abnormalities.

Toxicity: Vitamin E is relatively nontoxic. However, large intakes of vitamin E might interfere with absorption of Vitamin A and vitamin K. Doses necessary to elicit Toxicity far exceed those necessary for nutritional sufficiency.

Occurrence and properties: The richest sources of vitamin E are vegetable oils—including soy bean, corn, cottonseed, and safflower—and products made from these oils such as margarine, shortening, and mayonnaise, as well as wheat germ, nuts, and other grains. Meats, fish, animal fat, and most fruits and vegetables contain little vitamin E.

The acetate ester of α-tocopherol, rather than the free alcohol, is used as a food supplement because of its greater stability. Vitamin E is a slightly viscous, pale yellow, oily liquid obtained from molecular distillation of byproducts from vegetable oil refining or by chemical synthesis. The free alcohol form is highly unstable to oxidation and is used in foods as an antioxidant to stabilize the lipid component of foods. Cold water soluble forms have been produced by encapsulation with a suitable matrix.

Commercial forms: Some of the forms available as nutrients and/or dietary supplements are DL-α-Tocopherol, D-α-Tocopherol concentrate, Tocopherols concentrate, mixed, D-α-Tocopheryl acetate, DL- α-Tocopheryl acetate, D-α-Tocopheryl acetate concentrate and D-α-Tocopheryl acid succinate.

Vitamin K

Vitamin K functions in the liver as an essential cofactor for carboxylase. This enzyme converts specific glutamic acid residues of precursor proteins to a new amino acid, gammacarboxyglutamic acid (Gla). Vitamin K–dependent blood clotting factor prothrombin (factor II) and factor VII, IX, and X are included in the proteins. Vitamin K plays a crucial role in blood clotting activities. Vitamin K is not only associated with coagulation, but also with additional functions in bone, kidney and possibly other tissues.

Deficiencies in vitamin K status are relatively uncommon. However, since newborns do not have the necessary menaquinones, due to poor vitamin K placenta transfer, administration of vitamin K to newborns has become common practice.

Toxicity: There is no known Toxicity associated with the administration of high doses of vitamin K1. However, administration of vitamin K3 to infants is associated with hemolytic anemia and liver Toxicity.

Occurrence and properties: Green leafy vegetables, especially broccoli, cabbage, turnip greens, and lettuce, have large amounts of vitamin K. Other vegetables, fruits, cereals, dairy products, eggs, and meat contain smaller amounts. A significant amount of vitamin K is formed by the bacterial flora of the human lower intestinal tract. Vitamin K is fairly resistant to heat. Compounds tend to be unstable in the presence of alkali and light.

Commercial forms: Vitamin K is a clear, yellow to amber, very viscous liquid that is stable in air but decomposes when exposed to sunlight. It is slightly soluble in alcohol, and soluble in dehydrated alcohol, in chloroform, in either, and in vegetable oils. It is insoluble in water.

Vitamin C

Vitamin C is the antiscorbutic vitamin. First described during the Crusades and a common plague of early explorers and voyagers, scurvy is now virtually eliminated with the presence of vitamin C in adequate diets. Vitamin C, or ascorbic acid, has many functions as either a coenzyme or cofactor. Because of its ease of losing or taking on hydrogen, it is essential in metabolism. It is well recognized for its role in enhancing the absorption of iron. Vitamin C maintains the integrity of blood vessels. Vitamin C also promotes resistance to infection.

Toxicity: Since vitamin C is one of the most commonly used supplements, excessive intake may occur. Symptoms may include diarrhea, abdominal bloating, iron over-absorption, nausea, and kidney stones.

Occurrence and properties: The best sources of vitamin C are fruits and vegetables, preferably acidic and fresh. These sources include citrus fruits, raw leafy vegetables, tomatoes, broccoli, strawberries, cantaloupe, cabbage, and green peppers. Content of ascorbic acid in fruits and vegetables varies with the conditions under which they are grown and degree of ripeness when harvested. Ascorbic acid is a stable, odorless, white solid which is soluble in water, slightly soluble in alcohol, and insoluble in organic solvents.

Commercial forms: l-Ascorbic acid is white or slightly yellow crystals or powder. It will gradually darken on exposure to light, and is reasonably stable in air when dry but will rapidly deteriorate in solution in the presence of air. It has to be stored in tight light-resistant containers. Ascorbate is available in vitamin tablets and many multivitamin supplements contain ascorbate.

Thiamin

Also known as vitamin B1, thiamin has been known as the vitamin which prevents beriberi. Thiamin is essential in energy transformation and membrane and nerve conduction as well as in the synthesis of pentoses and the reduced coenzyme form of niacin. First identified in 1897 it was not until 1936 that the vitamin was synthesized. There are two major forms, thiamine pyrophosphate (TPP) and thiamin triphosphate (TTP). In the TPP or diphosphate form it is required for the oxidative decarboxylation of pyruvate to acetyl CoA, providing entry of oxidizable substrates into the Krebs cycle for energy. Thiamin is most strongly linked with carbohydrate metabolism even though it is also needed to metabolize fats, proteins, and nucleic acids.

Severe thiamin deficiency in humans leads to impairment of the auditory and visual pathways in the brain stem as well as impairment of the endocrine pancreas and the heart. In addition, clinical features of thiamin deficiency experienced by the chronic alcoholic includes ataxia and altered memory and alcoholic peripheral neuropathy.

Toxicity: Even in high oral doses, thiamin has been found to have no toxic effect other than possibly gastric upset.

Occurrence and properties: Food sources include meat, fish, whole cereal grains, fortified cereal and bakery products, nuts, legumes, eggs, yeast, fruits, and vegetables.

Thiamin esters are relatively stable in the dried state if stored at low temperatures in the dark. In solution thiamin is stable at pH 2–4, and Thiamin pyrophosphate (TPP) is stable at pH 2–6 at low temperature. All thiamin vitamers are unstable at elevated temperatures and under alkaline conditions.

Commercial forms: Thiamin mononitrate is available as white to yellowish white crystals, or crystalline powder, usually having a slight characteristic odor. It is slightly soluble in alcohol and in chloroform. Thiamin mononitrate should be stored in tight, light-resistant containers. Thiamin hydrochloride is also available as white to yellowish white crystals or crystalline powder.

Riboflavin

Deficiencies of riboflavin are usually in combination with deficiencies of other water soluble vitamins. Intake of riboflavin must be low for several months in order for signs of deficiency to develop. Pellagra like symptoms including skin lesions around the mouth, nose, and ears occur when there is a riboflavin deficiency.

Toxicity: There are no known effects from Toxicity.

Occurrence and properties: Riboflavin is widely distributed in small amounts. The best sources are milk, cheddar cheese, and cottage cheese. Other good sources include eggs, lean meats, broccoli, and enriched breads and cereals.

Vitamin B2 is heat-, oxidation-, and acid-stable but is sparingly soluble in water and disintegrates in the presence of alkali or light, especially ultraviolet. Very little riboflavin is lost in cooking and processing of foods.

Commercial forms: The commercial form is a yellow to orange yellow crystalline powder having a slight odor. It melts at about 280oC with decomposition, and its saturated solution is neutral to litmus. When dry it is not affected by diffused light, but in solution light induces deterioration. It is less soluble in alcohol than in water. In ether and in chloroform it is insoluble but very soluble in dilute solutions of alkalies. It should be stored in tight, light-resistant containers.

Niacin

Niacin is a generic term used for nicotinic acid and its derivatives that exhibit the biological activity of niacinamide. These function as a component of the coenzymes nicotinamide adenine dinucleotide (NAD) and nicotinamide adenine dinucleotide phosphate (NADP), which are present in all cells. These coenzymes are essential in oxidation–reduction reactions involved in the release of energy from carbohydrates, fats, and proteins. The coenzyme NAD is also used in glycogen synthesis. Niacin was identified as a result of the search for the cause and cure of pellagra. Clinical signs of pellagra are commonly referred to as the 3 Ds; diarrhea, dementia, and dermatitis. The most characteristic sign is a pigmented rash that develops symmetrically in areas of the skin exposed to sunlight.

Toxicity: High doses of niacin may be toxic to the liver. Large doses of nicotinic acid have been shown to reduce serum cholesterol concentrations. However, there are side effects that include flushing of the skin, hyperuricemia, hepatic and ocular abnormalities, and occasional hyperglycemia.

Occurrence and properties: Niacin is widely distributed in plant and animal foods. Good sources include meats (including liver), dairy products, cereals, legumes, and seeds. Green leafy vegetables and fish, especially shellfish, as well as enriched breads and cereals, also contain appreciable amounts.

Niacin in general is one of the more stable vitamins. Free forms of the vitamin are white stable crystalline solids. Nicotinamide is more soluble in water, alcohol, and ether than is nicotinic acid. Nicotinic acid's stability is pH independent.

Vitamin B6

There are six nutritionally active B6 vitamers occurring in foods. Its original form is pyridoxine (PN), pyridoxal (PL), pyridoxamine (PM), and the corresponding 5′-phosphate esters pyridoxine phosphate (PNP), pyridoxal phosphate (PLP), and pyridoxamine phosphate (PMP). The esters PLP and PMP function primarily in transamination and other reactions related to protein metabolism. The former is also necessary for the formation of the precursor of heme in hemoglobin. Vitamin B6 is essential in the metabolism of tryptophan and its conversion to niacin. The immune responses of young and elderly humans are also affected by the nutritional status of vitamin B6. Signs of deficiency in adults include stomatitis, cheilosis, glossitis, irritability, depression, and confusion.

Occurrence and properties: Best sources of vitamin B6 are yeast, wheat germ, pork, liver, whole grain cereals, legumes, potatoes, bananas, and oatmeal.

Vitamin B6 is considered relatively labile with its degree of lability influenced by pH. All forms are relatively heat stable in acid medium and are heat labile under alkaline conditions. In aqueous solution, most forms are light sensitive.

Commercial forms: Vitamin B6 is available commercially in the form of pyridoxine hydrochloride as a colorless or white crystal or a white crystalline powder. It is stable in air and slowly affected by sunlight.

Pantothenic Acid

Pantothenic acid was first synthesized in 1940. As a primary constituent of coenzyme A, it is involved in many areas of cellular metabolism including the synthesis of cholesterol, steroid hormones, vitamin A, vitamin D, and heme A. It is involved in the release of energy from carbohydrate and in the degradation and metabolism of fatty acids.

Pantothenic acid is widely available so deficiencies are rare. When there are conditions of severe malnutrition, deficiencies may be detected that mainly affect the adrenal cortex, nervous system, skin, and hair. Insufficiency is characterized by a burning sensation in the soles of feet.

Occurrence and properties: Pantothenic acid is widely distributed in nature. Rich dietary sources include meat, fish, poultry, eggs, whole grain products, and legumes.

Pantothenic acid is relatively stable at neutral pH. Strong acids, alkalis, or thermal processing will hydrolyze pantothenic acid. Calcium pantothenate occurs as a slightly hygroscopic, white powder. It is odorless, has a bitter taste, and is stable in air.

Commercial forms: Commercial available forms of pantothenic acid are Calcium pantothenate, racemic Calcium pantothenate, Calcium chloride double salt of DL- or D-calcium panthothenate, Dexpanthenol (dextrorotatory isomer of the alcohol analog of pantothenic acid and occurs as a clear, viscous, somewhat hygroscopic liquid having a slight odor) and DL-Panthenol.

Folate

Folate is a generic term for a group of compounds chemically and nutritionally similar to folic acid. Essential to cell division, folate plays an important role in the synthesis of the purines guanine and adenine and the pyrimidine thymine, which are compounds utilized in the formation of nucleoproteins deoxyribonucleic acid (DNA) and ribonucleic acid (RNA). Folate is also essential for the formation of both red and white blood cells and serves as a single carbon carrier in the formation of heme. Low levels of plasma folate and vitamin B12 lead to elevated plasma homocysteine, which has been associated with increased risks of coronary heart disease. Deficiency of folate results in poor growth, megaloblastic anemia, and other blood disorders elevated blood levels of homocysteine, glossitis, and gastrointestinal tract disturbances.

Occurrence and properties: Folates are ubiquitous in nature and are present in nearly all natural foods. However, they are highly susceptible to oxidation and folate content in foods and may be destroyed by protracted cooking or processing, such as canning. Best sources are legumes such as kidney beans, lima beans, lentils, fresh dark green leafy vegetables, especially spinach, asparagus, turnip greens, and broccoli.

Crystalline folic acid is yellow. The free acid is almost insoluble in cold water, and the disodium salt is more soluble. Folic acid is destroyed at a pH below 4, but is relatively stable above pH 5.

Commercial forms: Folic acid is a yellow or yellowish orange, odorless crystal or crystalline powder. It is insoluble in acetone, in alcohol, in chloroform, and in ether, but dissolves in solutions of alkali hydroxides and carbonates.

Vitamin B12

Initially vitamin B12, also known as cobalamin, was identified as the extrinsic factor of food that is effective in the treatment of pernicious anemia. Vitamin B12 is synthesized by bacteria. It is essential for normal function in the metabolism of all cells, especially cells of the gastrointestinal tract, bone marrow, and nervous tissue. It participates, along with folic acid, choline, and methionine, in the transfer of methyl groups in the synthesis of nucleic acids, purines, and pyrimidine intermediates. Vitamin B12 and folate are closely related with each depending on the other for activation. With vitamin B12 deficiencies, impaired DNA synthesis results leading to megaloblastic anemia, glossitis, hypospermia, and gastrointestional disorders. Lack of vitamin B12 also results in subacute degeneration of cerebral white matter, optic nerves, spinal cord, and peripheral nerves with resulting symptoms of numbness, tingling, burning of the feet, stiffness, and generalized weakness of the legs.

Toxicity: No toxic effects are known.

Occurrence and properties: Vitamin B12 is found almost exclusively in foods derived from animals. Usual dietary sources are meat and meat products and to a lesser extent milk and milk products. Richest sources are liver and kidney. Plant products have very little vitamin B12. Fermented soy products and some sea algae do not provide an active form of vitamin B12. Megadoses of vitamin C may adversely affect the availability of vitamin B12 from Food.

Cyanocobalamin crystals are dark red, and the substance absorbs water. Cobalamins are destroyed by heavy metals and strong oxidizing or reducing agents. Aqueous solutions are neutral with maximum stability at pH 4.5 to 5.0.

Commercial forms: Vitamin B12/cyanocobalamin are dark red crystals or amorphous or crystalline powder. In anhydrous form it is very hygroscopic and when exposed to air it may absorb about 12% of water. Packaging and storage should be in well-closed containers.

Biotin

Biotin was first isolated in 1936 and synthesized in 1943. First named vitamin H, it was later proved to be the same as a potent growth factor in yeast known as coenzyme R and was renamed biotin. It functions as the coenzyme for reactions involving the addition or removal of carbon dioxide to or from active compounds. These coenzymes are involved in gluconeogenesis, synthesis, and oxidation of fatty acids, degradation of some amino acids, and purine synthesis. The absorption of both dietary biotin and any biotin synthesized by intestinal bacteria is thought to be prevented by dietary avidin, a glycoprotein found in raw egg white. Thus the name "egg-white injury" given to the syndrome of severe dermatitis, hair loss, and neuromuscular dysfunction.

Occurence and Properties: Biotin is widely distributed with major food sources including liver, rice, egg yolks, and vegetables. It may also be found in seeds of many cereals and oilseeds, however it is largely unavailable in this form. Avidin reduces the bioavailability of biotin. Biotin will occur in food as free biotin but is usually bound to protein.

Biotin is soluble in water. It is more soluble in hot water and in dilute alkali and four times more soluble in 95% ethanol than in cold water. It is not soluble in other organic solvents. Biotin does not fluoresce nor is it a UV absorber.

Commercial forms: cis-Hexahydro-2-oxo-1H-thieno[3,4]imidazole-4-valeric acid and d-biotin is a practically white crystalline powder. It is stable to air and heat. Packaging and storage should be done in tight containers.

Proteins as Nutritional Additives

Dairy ingredients can boost the protein content of foods, while delivering on the promise of superior flavor with minimum guilt. Diets higher in protein have been associated with increased satiety, so foods fortified with dairy proteins show great promise in foods intended for weight management market. Protein products high in casein independently or in combination with whey proteins are not new. Traditional procedures for manufacture of such products include acid precipitation or rennet induced coagulation. To achieve proper solubility and functionality, the precipitated casein is frequently sold in the form of sodium or calcium caseinates. If whey proteins are to be included with casein, co-precipitation techniques are required (i.e. calcium co- precipitates). Vegetable proteins are also good sources of plant proteins. The most successful one being the soy protein products.

Protein Sources and their Digestibility

Some of the protein ingredients obtained from various sources are tabulated in table.

Table: Commercial protein ingredients used for food manufacture.

Source	Protein ingredients
Soy, wheat, maize and other vegetable sources	Grits, flour, concentrates or isolates, fractionated proteins.
Milk	NFDM, acid casein, rennet casein, caseinate, Co-precipitate, WPC, MPC, Lactalbumin.
Egg	Liquid egg white, dried egg, egg yolk.
Meat	Meat hydrolysates, comminuted meat, mechanically recovered protein, gelatin.
Fish	Fish protein hydrolysates, collagen, muscle.

NFDM – Nonfat dried milk, WPC – Whey protein concentrate, MPC – Milk protein concentrate.

In order to confer the benefits derived out of protein, it has to be digested easily. The protein quality can be determined by finding the content of the first limiting essential amino acid of the test protein as a percentage of the content of the same amino acid content in a reference pattern of essential amino acids. This in turn can be corrected to have true fecal digestibility of the test protein. The value obtained was referred to as the protein digestibility corrected amino acid score (PDCAAS) which is the preferred method for measurement of the protein value in human nutrition. The protein digestibility corrected amino acid score of various proteins is shown in table.

Table: Comparison of selected protein products for human nutrition.

Type of protein	Protein digestibility corrected amino acid score (PDCAAS)
Casein	1.0
Egg white	1.0
Soy protein concentrate	0.99
Soy protein isolate	0.92
Pea flour	0.69
Peanut meal	0.52
Rice	0.47
Whole wheat	0.42
Wheat gluten	0.25

Dairy Proteins

The two dairy proteins are caseins (80% of total protein) and whey proteins (20% of total protein). Whey proteins remains soluble over a wide range of pH levels and have superior gelling and water binding capability, unless they are heated and become denatured. Dairy proteins with their hydrophilic and hydrophobic polar nature function as emulsifiers and can be used to replace fat, improve texture and shelf life of a variety of breads, meats and frozen desserts. Whey proteins can be used to boost the protein content of a wide variety of beverages. They can be used in meal replacers, smoothies, isotonics, protein waters and sports drinks.

Traditionally Used Milk Derived High Protein Powders

The examples of traditional high protein powders of dairy origin are acid casein, rennet casein, calcium co-precipitate, and derivatives of casein viz., sodium or calcium caseinates.

- Rennet and acid caseins: These are precipitation of a casein curd from skim milk (acid and rennet enzyme for acid and rennet casein respectively) followed by washing and drying. The whey proteins are lost in preparing such products. They are approximately 90% protein on dry basis.
- Caseinates: These are made by treating the rennet or acid casein with alkali (sodium or calcium salt) and then drying. The products are Na-caseinate and Ca-caseinate. The products contain approximately 90% protein. This improves the solubility of the caseins for use as an ingredient in foods.
- Co-precipitates: These are prepared by adding calcium chloride or dilute acid to skim milk followed by a heating step to cause a curd formation that captures both the caseins and whey proteins. The products are approximately 90% protein.
- Milk protein blends: Combinations of dry milk derived products from whey, casein, co-precipitates, caseins, nonfat dry milk, etc. technically can be blended together to produce a milk protein concentrate. They can be blended to produce products with a wide range of compositions for use as custom ingredients in food product manufacture. However, the functionality (i.e., solubility, flavor, etc.) may be very different than Milk Protein Concentrate (MPC) prepared by ultrafiltration of milk.

Functional Properties of Proteins

Functional properties of proteins could be classified according to the mechanism of action on three main groups: (i) properties related to hydration (absorption of water/oil, solubility, thickening, wettability) (ii) properties related to protein structure and rheological characteristics (viscosity, elasticity, adhesiveness, aggregation and gelification), and (iii) properties related to protein surface (emulsifying and foaming activities, formation of protein-lipid films, whippability). Some of the functionalities of protein ingredient in food product is tabulated in Table.

Table: Multiple protein functionality in selected processed foods.

Type of food	Multiple functionality
Beverage	Solubility, colour, grittiness
Baked goods	Emulsification, foaming, gelation,
Dairy substitutes	Gelation, foaming, emulsification
Egg substitutes	Foaming, gelation
Meat emulsions	Emulsification, foaming, gelation, adhesion-cohesion
Soups and gravies	Viscosity, emulsification, water adsorption
Topping	Foaming, emulsification
Whipped dessert	Foaming, gelation, emulsification

Now, with the adoption of newer technologies such as ultrafiltration, microfiltration and nanofiltration, purified protein products having a wide range of functionality can be tailor made. Protein powders based on such novel technologies include Whey Protein concentrates (WPC), Milk Protein Concentrates (MPC), Micellar casein powder, etc.

Milk Protein Concentrates

MPC production begins with ultra-filtration of skim milk to make a lactose-reduced, skim milk concentrate, which is further concentrated in an evaporator by using two-stage drying process and, finally, spray dried into powder form. Milk protein concentrate contains high levels of protein and low lactose. Milk protein concentrate preserves the unique casein to whey ratio in milk. Milk protein concentrates (MPCs) are complete dairy proteins (containing both caseins and whey proteins) that are available in protein content ranging from 40.0 to 89.0 percent. As the protein content of MPCs increases, the lactose levels decreases.

Table: Proximate composition of some high protein dairy based food ingredients.

Constituents	MPC-42	MPC-80	WPC-34	WPC-80	Acid casein*	Rennet casein*
Moisture	5.0	5.0	4.0	5.0	10.0	10.0
Fat	2.0	3.0	5.0	5.0	2.0	2.0
Protein	42.0	80.0	34.0	80.0	90.0	84.0
Lactose	46.0	6.5	50.0	6.5	1.0	1.0
Ash	5.0	5.5	7.0	3.5	2.3	7.6

* On dry matter basis, except for moisture.

Typical MPCs offered are MPC42, MPC70, MPC75, MPC80, MPC85 and milk protein isolate (MPI with 90% protein). MPCs are produced by ultrafiltration (UF) or by blending different dairy ingredients. Although UF is the preferred method for producing MPCs, they also can be produced by precipitating the proteins out of milk or by dry-blending the milk proteins with other milk components. The functionality of MPCs may differ depending on how they were produced. For comparison, skim milk powder (SMP) contains about 35.0% protein and 52.0% lactose.

Functional Characteristics of MPCs

Some of the common functional characteristics that MPCs offer in various food and beverage applications include:

- Solubility: Ultrafiltration (UF) preserves the original protein structure of MPCs, which results in a greater solubility. Since they are readily soluble, MPCs are beneficial in dry dairy-based mixes. MPCs that are stored at ambient to cool temperatures (40 to 70°F) and low humidity retain good solubility for 6-8 months. Dissolvability of MPCs can be improved by increasing the sodium or potassium content, which makes MPCs more suitable for use in specific food and beverage applications such as meal replacement beverages and process cheese.
- Emulsification/Foaming/Whipping: The proteins in MPC act at the oil/water interface to form and stabilize fat emulsions in sausages and other processed meats, dairy drinks, soups, vinaigrettes, sauces and bakery products. The proteins in MPC act at the air/water interface

to form a stable film of air bubbles. Hence during its application, it stabilizes meringues, mousses, cake, ice creams, whipped cream. The whey protein denaturation levels, different types of casein-whey protein complexes, various levels of casein-bound calcium and soluble casein, and manipulating casein micelle size can help in improveing MPC functionality. This would enhance the emulsions (oil/water interface) and whipping (air/water interface) of MPCs for application in sausages, processed meats, dairy drinks, soups, vinaigrettes, sauces and bakery products. MPC can form a consistent film of air bubbles to stabilize meringues, mousses, cakes, ice creams, whipped cream and soufflés.

- Water Binding/Viscosity: Efforts are underway to enhance the functionality of MPCs to improve water binding and viscosity similar to rennet casein utilizing enzymatic process before spray-drying. The functionalized MPCs will enhance the viscosity of the food product and will be applicable in products such as cheeses, firm yogurts, sauces, milk-based drinks and creams for desserts and bakery products.

Application of MPCs

MPCs are used for their nutritional and functional properties. MPC averages approximately 365 kcal/100g. Higher-protein MPCs provide protein enhancement and a clean dairy flavor without adding significant levels of lactose to food and beverage formulations. MPCs also contribute valuable minerals such as calcium, magnesium and phosphorus to formulations, which may reduce the need for additional sources of these minerals. MPCs are multifunctional ingredients and provide benefits such as water binding, gelling, foaming, emulsification and heat stability.

MPCs are currently used for manufacturing products like process cheese, cream cheese, ice cream, yogurt/fermented dairy and meal replacement beverages. Typically, lower-protein MPCs are used as ingredients in cheese applications, while higher-protein MPCs are used in beverage and bar applications. They are used as an ingredient in applications that rely on nonfat dry milk (NFDM) but require a higher level of protein where casein and caseinate are used traditionally. Use of MPCs as an ingredient is growing at the expense of casein and caseinate, due to their improved flavor profile. Common applications include infant formulas, desserts, baked goods, toppings, low-fat spreads, dairy-based dry mixes, dairy-based beverages, sports and nutrition beverages/foods, geriatric nutritional products, weight loss beverages/foods and some process cheese products. MPCs are now widely used in a variety of protein-enhanced foods, but primarily in meal replacements, nutritional beverages and bars.

MPCs can be used to boost the protein content and reduce the sugar levels of a wide variety of puddings, nutritional beverages, process cheeses and frozen desserts.

- Yoghurt: Skim milk powder (SMP) is used commonly to enrich yogurt at a rate of 3.0-4.0% to increase the TS. Addition of milk protein helps form a firmer body and reduces whey expulsion in the final product. The dry matter content of typical commercial yogurts ranges from 13.0- 17.0% because of enrichment of the base milk. There is a limit to the quantity of SMP that can be used to get firmer body because large amount can produce powdery mouth feel, and excess lactose from SMP can lead to extreme acidity during storage of the final product. MPC possesses unique casein to whey protein ratio. Such qualities make MPC an alternative to NFDM in yogurt.

- Pizza cheese: The addition of MPC (to standardize cheese milk to protein/fat of 1.40) improved the yield from 10.34% in control pizza cheese to 14.50% and 16.65%, respectively, for pizza cheese made by starter culture (SC) and direct acidification (DA) methods. The TS recoveries were in the order, SC > DA > control. DA pizza cheese had the best meltability.

Micellar Casein Concentrates

Manufacture of Micellar Casein Concentrates (MCC) from milk by microfiltration (MF) produces a range of compositions depending on the amount of milk derived whey (MDW) protein removed. Further concentration and diafiltration (DF) can increase the total protein and decrease the amount of lactose in the final ingredient. MCC ingredient provides notable nutritional benefits and is an excellent source of all essential amino acids and calcium. Apart from nutritional benefit, it offers a unique functional benefit for application in retort beverages and cheese production. An MCC with 95.0% of the MDW protein removed has approximately 84.0% protein, of which 80.5% was casein and 3.7% was whey protein.

Table: Composition of Micellar casein concentrate as against caseinate powder.

Constituents	Micellar casein concentrate (MCC84)*	Caseinate
Moisture	5.20	8.0
Fat	2.20	2.0
Protein	84.10	88.0
Lactose	0.70	1.0
Ash	7.80	-
pH	-	8.0

* Values are on dry matter basis.

The composition of MDW protein ingredients are unique compared with those derived from cheese whey. The protein composition of cheese whey differs from MDW because it contains glycomacropeptide (GMP), which is cleaved from κ-casein through the action of chymosin in cheese making. The microfiltration process retains milk fat in the retentate, which yields an MDW that is essentially free of fat (< 0.3%), even after further concentration to 80.0% protein. Alternatively, cheese-whey-derived WPCs (CD-WPC) with 80.0% protein typically contain 6.0- 7.0% fat. This difference in fat content also is a major factor in explaining their differences in functionality, flavor and appearance.

The functional properties of MDW protein concentrates (MD-WPC) are typically better than CD-WPC. MD-WPC has greater foaming, gel strength, solubility and emulsification capacity than CD-WPC. Owing to the higher denaturation levels in the commercial CD-WPC, it has greater water-holding capacity (WHC) and lower solubility.

Sensory properties and appearance: The sensory properties of MD-WPC are notably different from commercial CD-WPC. Similar aroma compounds have been identified both protein products, many of which are lipid and protein oxidation products. Commercial CD-WPCs have higher concentrations of volatile compounds such as hexanal, heptanal and pentanal compared with MD-WPC. Besides a cleaner flavor, MD-WPC has a very different appearance in solution compared with a CD-WPC, which is related to the fat content. The low fat content in MD-WPC gives it a clear appearance in solution, while a CD-WPC appears very milky and cloudy. This clear appearance allows MD-WPC at any protein level to be used in clear beverage applications.

Specific Applications of Micellar Casein Powder

- Beverages: MCC would be a good choice for neutral pH UHT or retort-processed nutritional drinks because of their heat stability. The use of MD-WPC produces a clear, high-acid beverage (pH 3.4) that has similar heat stability and clarity to a whey protein isolate (WPI). No significant change in pH, viscosity or solubility occurred during the shelf life (39 weeks at 20°C) of either beverage. The WPI beverage developed some off-flavors typically associated with high-protein beverages, while the MD-WPC-80 beverage had a clean flavor and aroma when compared with WPI beverages.
- Cheese: The practice of cheese milk standardization is common and is typically done to improve yield and produce cheese with a consistent composition. Cheese makers in the United States are allowed to use ingredients such as condensed skim milk and NFDM for this purpose in standard of identity cheeses, such as Cheddar and Mozzarella. The ideal ingredient for standardization of cheese milk would contain predominantly casein protein, as that is the primary protein in cheese. Use of microfiltration retentates (liquid MCC) in the manufacture of cheese revealed that MCC is a useful ingredient for increasing production capacity while maintaining cheese quality. Fractionation of milk by microfiltration (MF) creates more value-added, milk-derived ingredient opportunities that have diverse applications for use by the dairy and food industries. These ingredients can provide unique functionalities along with clean flavor and excellent nutrition.

Whey Protein Concentrates and Isolates

About half of whey proteins are β-lactoglobulins with (Isoelectric pH (IEP) of ~5.4); about 20.0% are α-lactalbumin (IEP of 4.4) and a little more than 10.0% are glycomacropeptides (IEP of < 3.8). Whey protein concentrates (WPC) is prepared by passing whey (cheese) through ultrafiltration membrane to recovery whey proteins in the retentate. Ultrafiltration retains (retentate) any insoluble material or solutes > 20,000 Da molecular weight. The rest of the whey stream passes through the membrane, driven by the applied pressure and is called permeate. The permeate contains most of the lactose, minerals and water from the whey. The retentate, the volume of which is about 1-4% that of the feed whey, is spray dried to a powder containing 35- 85% protein as desired. Usually two types of WPCs are available commercially viz., WPC-35 and WPC-70; the number indicating the protein content of powders.

Functionality of Whey Protein Powders

- Clarity/Opacity: If clarity is the goal, then the ingredient of choice is WPI, which has over 90% protein and very low fat. Isolates also work well in opaque beverages, but in these applications, an MPC70 or WPC80 maybe more cost-effective and provide good opacity.
- Solubility: In protein waters and isotonics, whey proteins are easily soluble at a pH of 3.4, where soy and casein would precipitate out. Whey proteins also work well at neutral pH (i.e. 7.0).

- Processing: Other factors that affect the appearance and stability of whey protein beverages include ionic strength, hydration time, sugar and water. Adequate hydration can reduce turbidity by as much as 50%.
- Structure and browning: If goal is to improve grain structure, retain moisture, enhance crust browning and improve flavor then high heat NFDM, WPC34 or specialty dairy blend might be a good choice.

Applications of Whey Protein Concentrates

WPC and Whey protein isolates (WPI), for example, are added to foods and beverages for both nutritional and functional purposes.

Infant formula: Cheese-derived whey proteins have been the predominant source of whey protein for infant formulas today. More recently, milk-derived WPC or WPI has been highlighted to infant formula manufacturers as a source of protein not containing glycomacropeptide, which is absent in human milk.

Whey ingredients can be used in processed meats as partial replacements for meat, as binders, as flavor enhancers and meat analogues. Whey ingredients can improve yield and prevent moisture loss in ham and surimi. In meat applications, the binding of myofibrillar meat proteins decreases during cooking, due to protein denaturation. Under appropriate heating conditions, whey proteins can form irreversible gels by restructuring into extended, 3-dimensional networks.

The water binding capability of whey proteins makes them useful in a wide variety of soups and sauces. For products, that will be exposed to high temperatures, either an MPC or heat stable whey proteins may be needed.

Cheesemaking: Incorporation of WPC led to an increase in the yield of Cheddar cheese from 9.96% (cheese prepared through traditional process) to 10.31% (WPC added after whey extracting) and 11.21% (WPC added during inoculation with starter culture).

Vegetable Protein Concentrates and Isolates

Commercially, the extraction and purification of vegetable protein can be performed by processing in neutral or in acid medium to produce concentrates (48–70% protein) or in alkaline medium leading to an isolate as final product (85–90% protein). Protein concentrates result from removing non-proteic constituents, mainly soluble minerals, carbohydrates, low molecular weight nitrogen compounds and antinutritive factors from full fat or, more usually, from defatted meals.

Soy Protein as High Protein Ingredient

Since 1960s, soy protein products have been used as nutritional and functional food ingredients in every food category available to the consumers. Protein supplementation of cereals is desirable in many instances because cereals have low protein content and are imbalanced in essential amino acids (EAA) composition. Soy protein products are an ideal source of some of the EAAs used to complement cereal proteins.

Forms of Soybean Protein

At present, the various forms of soybean protein are utilized primarily for their functional effects rather than for their nutritional properties. After oil removal from soybeans, the remaining proteinaceous material is referred to as soybean defatted flakes. Three forms of soyprotein products exist as protein supplements or as protein source and their content ranges from 50% to > 90%, namely soy flours/grits, soy protein concentrates and soy protein isolates. The protein content of various soybean protein ingredients are depicted in table.

Table: Composition of various protein ingredients of soybean origin.

Protein ingredients	Moisture	Protein	Fat
Flours and grits Full fat Defatted	6.00 6.00	41.00 53.00	20.50 0.60
Concentrate	6.70	66.20	0.30
Isolate	4.70	92.80	< 0.10
Texturized soy protein	9.00	50.00	6.0-8.0

Soy Protein Concentrates

Soybean protein concentrates (SPC) contain ≥ 70.0% protein on moisture-free basis. Protein concentrates are prepared from defatted flakes or flour by removing the oligosaccharides, part of the ash, and some of the minor components. SPC preparation involves retaining the soybean globulin proteins while selectively removing the soluble sugar carbohydrates. The physical properties of SPC will differ with based on the method of preparation (i.e. heat, acid leach or alcohol leach). Concentrates have a reduced flavor level as compared to flours and grits.

SPC were developed as flavor improvers and for increasing the protein content of foods. These products provide high quality protein and are comparable to animal protein. The Food and Drug Administration (FDA) has allowed a claim that "A daily diet of 25 g of soybean protein which is low in cholesterol and unsaturated fat can reduce total LDL cholesterol moderately" (FDA, 1999). SPC has improved flavor characteristics and provides functional characteristics such as fat-micelle stabilization, water and fat absorption, viscosity control, and texture control in forming fat emulsions in food systems. The substitution of flour by 5.0% SPC increased the blend protein content by approximately 2.5%.

Since the neutralized acid leach and steam injection/jet cooking processes can result in a product that has high dispersibility, concentrates of these types will have more desirable functional properties in emulsion type applications. These dispersible functional concentrates have many of the properties exhibited by neutralized isolates. Concentrates may vary as to color, flavor, particle size, and water and fat absorption. All of the SPCs, regardless of the process used, do have fat and water holding properties (partially due to their polysaccharide content), and modify viscosity and textural characteristics of the food system. The alcohol-denatured concentrates exhibit good nutritional value and are used in many applications requiring protein fortification. They may also be texturized.

Soy Protein Isolates

The most refined form of soybean proteins are the soy protein isolates (SPI), which contain ≥ 90.0% protein. They are prepared by removing the water-insoluble polysaccharides, as well as the oligosaccharides and other low-molecular-weight components that are separated. Defatted flakes or flours, which have received a minimum of moist heat treatment, are extracted with water plus alkali at a pH of 7.0-8.5. The insoluble residue, containing the water-insoluble polysaccharides plus residual protein, is then separated. The clarified extract containing the bulk of the proteins plus the sugars is adjusted to about pH 4.5. These precipitated proteins are removed by centrifugation or filtration. These are then washed and dried to give the isoelectric protein. The protein is neutralized before drying.

Functional Properties of Soy Proteins

SPC and SPI provide highly concentrated protein sources, high lysine content, bland flavor, and reduction of flatulence factors and reducing sugars, and they may lead to improved overall product quality.

The common functionalities of soy protein products include solubility, water absorption and binding, viscosity control, gelation, cohesion, adhesion, elasticity, emulsification, fat absorption or repulsion, flavor binding, foaming, whipping, and color control.

Commercially available soy proteins exhibit a wide range of solubilities at pH 7 ranging from 25.0 to 95.0%; solubility being dependent on pH and salt concentration. Higher pH results in high solubility. Commercial soy proteins also have a wide range of water binding ability. Insoluble protein granules bind much more water than soluble proteins. Slurries of soy proteins have relatively low apparent viscosity and do not form strong gel networks; however, heat or alkali treatments can increase viscosity and improve gelation functionality.

Several SPIs have been developed for providing different functional or physical properties to meet the requirements of various food systems. The solubility of isolates ranges from 25.0 to 95.0 NSI (nitrogen solubility index). Other than solubility, emulsifying capacity, gelling, fat and water absorption, and viscosity should also be considered. The emulsion capacity of isolates can differ by a factor of nearly 4. Emulsion capacity or the ability of a protein to emulsify fat is an important functional parameter in many food products. Isolates can emulsify from 10 to about 35 mL of oil per 100 mg of protein. The film-forming properties of SPI are useful in certain meat products.

The application of heat and shear to the protein solutions can also alter their viscosity. Water and fat absorption properties are also utilized in meat and baking applications. Isolates have water absorption values from about 150 to 400%. Neutralized isolates are usually highly dispersible and gels under appropriate aqueous conditions. They possess both emulsifying and emulsion stabilizing properties, and are excellent binders of fat and water, and are good adhesive agents too. Hence, they are used in processed meat products, both coarse and fine emulsions (as in patties, loaves, and sausages). Isolates and concentrates can also be structured to have a fibrous appearance. These products are designed mainly for poultry and meat protein replacement. The fiber-like structure, adds texture and mouth feel to poultry rolls.

Commercial soy proteins display a range of emulsion and foam properties. For commercial food applications these organoleptic problems can be avoided through careful control of food processing conditions.

Applications of Soy Protein Products

Full-fat and defatted soy flours have been used successfully as major ingredients in low-cost replacements for milk solids in beverages for human consumption in developing countries. All SPCs have much improved flavor characteristics as compared to commercially available soy flours. In general, the greatest potential areas are where casein and NFDM are used, such as emulsion-type meat products, bakery products, nutritional powder drinks, and soup bases. Baby foods, cereals, dry food mixes, milk replacers and snacks are few more examples where powdered SPC may be used.

SPI is used in the food industry for nutritional (increasing protein content), sensorial (better mouth feel, bland flavour) and functional reasons (for applications requiring emulsification, water and fat absorption and adhesive properties). In general, the greatest potential areas for use of SPC are where casein and NFDM are used, such as emulsion-type meat products, bakery products, nutritional powder drinks, and soup bases. Baby foods, cereals, dry food mixes, milk replacers, and snacks are just a few more examples where powdered SPCs may be used.

SPI are the most acceptable products for dairy applications because of their fine particle size and dispersibility. The functional properties of emulsification, emulsion stability, color, and flavor/odor are critical factors in dairy applications. A few, such as dry and liquid coffee whiteners, liquid whipped toppings, pre-whipped toppings, and toppings of other emulsified food items to replace Na-caseinate enjoy distribution in few countries. Dairy-type foods utilizing SPIs include imitation milks, convenience beverage powders, frozen desserts, sour cream, sour cream dips, and related cheese-like products. Soy protein ingredients are now being used in emulsion- type cheeses to replace as much as 50% of Na-caseinate in some products. Soy and milk protein blends (combined to offer protein content similar to that of milk) are sold as ingredients in bakery products, sauces, meat products, and various fabricated foods, as complete or partial replacement of NFDM. Soy protein formulas are recommended for those infants, and others, who are allergic to milk protein and those who have lactose intolerance or lactase deficiency.

Water and fat absorption properties of SPI are utilized in meat and baking applications. Isolates have water absorption values from about 150.0 to 400.0%. Neutralized isolates are usually highly dispersible and gels under appropriate aqueous conditions. Some of the general-purpose isolates (non-dispersible) have good nutritional properties and vary mainly in their gelling and viscosity characteristics.

Isolates and concentrates can also be structured to have a fibrous appearance. These products are designed mainly for poultry and meat protein replacement. The fiber-like structure, for example, adds texture and mouth feel to poultry rolls.

- Meat products: The major application of SPI in meat and related product is based on the use of texturized SPI, in one form or another, to replace meat. The neurtralized isolates possess both emulsifying and emulsion stabilizing properties, and are excellent binders of fat and water, and are good adhesive agents. Due to such inherent properties, they are used in processed meat products, both coarse and fine emulsions (as in patties, loaves, and sausages). SPI and proteinates are used for their moisture and fat binding properties and also used as emulsion stabilizers. Typical usage levels are 1.0 to 4.0% on prehydrated basis. The use of SPI in these products permits reducing the proportion of expensive meat in the formulation, without reducing the protein content or sacrificing eating quality.

- Seafood products: SPI find good application in fish sausage and surimi (washed, minced fish flesh) based restructured fish products in Japan.
- Cereal products: SPI is sometimes used instead of, or in combination with concentrates and soy flour, in the formulation of milk replacer mixtures in bakery products. SPI has been used for protein fortification of pasta and specialty bread. In these applications, the high protein content and blandness of SPI are clear advantages.
- Dairy-type products: SPI are used in non-dairy coffee whiteners, liquid whipped toppings, emulsified sour cream or cheese dressings, non-dairy frozen desserts, etc. The basis for these applications is, demand for non-dairy (all-vegetarian, cholesterol-free, allergen-free) food products, as well as economy.
    - Imitation cheeses have been produced from isolated soy proteins, with or without milk or whey components. The types of cheeses which can be produced include soft, semi-soft, surface- cultured (imitation Camembert) and ripened hard cheeses.
    - Whipped toppings: Milk protein has been replaced with SPI (0.4% w/v) leading to products with a overrun of 350-400% and excellent stability. Proteinase treated SPI had superior whipping functionality compared to egg white.
    - Infant formulas: Infant formulas where milk solids have been replaced by soy products are well established commercial products. SPI is the preferred soy ingredient, because of its blandness, absence of flatus-producing sugars and negligible fibre content. The principal market for these products is lactose-intolerant babies. However, soy protein based dietetic formulas are finding increasing use in geriatric and post-operative feeding as well as in weight reduction programs.

Generalized Applications of Protein Ingredients

It is recommended to use 2.0% of WPC34 or 0.4-0.75% of WPC80/WPI for fortification of milk solids content of yoghurt mix, along with some skim milk powder. Increasing levels beyond this may lead to graininess, lumpiness, and a short/brittle texture.

Specific Applications of High Protein Powders

Specialty breads: The protein content of ordinary white bread is 8.0 to 9.0%. Specialty breads can be made with 13.0 to 14.0% protein by incorporating soy flour, concentrate, or isolate in the formula, along with vital wheat gluten and, if necessary, a lipid emulsifier. Without surfactants, incorporation of high levels of soy flour depresses loaf volume and gives poor crumb characteristics. In cases where 12% of soy flour was used in non-standardized bread, the Protein Efficiency Ratio (PER) increased from 0.7 to 1.95. In addition to the improvement in protein quality, the protein content also increased by > 50.0% at such level of supplementation. Soy- fortified wheat flour has been used worldwide in mass feeding and school lunch programs.

Increasing substitution (20.0 to 100.0%) of egg yolk with WPC in Gelato (frozen delicacy) making, led to increased overrun, number of air cells formation and texture characteristics of the product; control Gelato used 4.5-9.0% of egg yolk. Practical applications of WPC involving interaction between water and protein include yoghurt, hard pack ice cream, low-fat ice cream, non-fat ice cream, soft ice cream, sour cream, and coffee whiteners. WPC80 is substantially cheaper than egg yolk. Use of WPI (95% protein) as part substitution of NFDM led to desirable properties in frozen product as against use of Na-/Ca-caseinate, rennet casein (all these with 95.0% protein) and WPC-36 (36% protein); sodium caseinate enhanced the overrun in ice cream.

Other Vegetable Protein Powders

The fortification of human food with oilseed proteins is expanding rapidly. Much of the success has been achieved with soy protein products, but peanut protein and cottonseed protein products have also been included in human food.

Peanut Protein Concentrate

Use of peanut protein concentrates (PPC) at levels higher than 10.0% reduced the loaf volume of bread significantly compared to that of bread baked with other protein sources. Cottonseed protein product was added to raise the bread protein by 30.0% as compared to bread baked with 100% wheat flour (from 10.0 to 13.0% protein by weight). Acceptable loaves were obtained with blends containing cottonseed protein concentrate, spray or freeze dried at near neutral pH. The quality of baked bread was similar to the one made using commercial soy flour.

Recent Developments in High Protein Powders

Spray dried ultrafiltered retentate powder: Ultrafiltration is both a fractionation as well as concentration process. Subjecting milk to ultrafiltration, casein as well as whey proteins are retained in the retentate and the total protein content is increased as permeate is diffused through the membrane. Such UF concentrate when spray dried yields high protein powder ingredient whose composition is shown in Table.

Table: Comparison of physico-chemical properties of ultrafitered retentate powder with skim milk powder.

Parameters	Spray dried Skim milk powder	Spray dried UF retentate powder (made from 4:1 RVC)
Moisture	3.60	3.90
Protein	35.70	64.40
Lactose	53.00	24.00
Fat	0.50	0.50
Ash	7.70	7.80
Density (packed)	0.375	0.493
Flowability (g/s)	1.25	1.22
Solubility index	0.03	1.60

Traditionally, non-fat yoghurt has poor body and frequently has excessive whey separation. To avoid this gelling and stabilizing agents are invariably used. Use of high milk protein powder, while adding protein to yoghurt, served as a stabilizer to improve the body and texture.

A high protein milk powder (HPMP) has been prepared using ultrafiltration with diafiltration of skim milk. The final protein powder had 5.33% moisture, 88.0% total protein, 2.27% fat, 0.74% lactose and 7.5% ash. Through such processing, it was possible to reduce lactose up to 99.7%. Specific application of such HPMP is in low-fat and non-fat yoghurts. Supplementing skim milk with HPMP up to 5.6% protein could produce low-fat yoghurt of good quality. The added protein helped to provide a firm body, without the use of stabilizers. Addition of 1.0% HPMP to cheese milk and fat standardization produces Gouda cheese with good sensory properties as well as increased cheese yield.

Other Promising Protein Powders

Some of the other promising protein sources include spray dried egg yolk powder (5.5% moisture, 32.9% protein); rice bran protein, rice protein concentrate, canola protein concentrate, pea protein concentrate, barley protein concentrate, lupin protein concentrate, etc. Peanut flour contains 47.0 to 55.0% protein. Peanut protein concentrate had crude protein content of 70.2% but it had an unbalanced amino acid pattern and was deficient in lysine, threonine, methionine and tryptophan.

References

- Chloride-in-diet, nutrition: mountsinai.org, Retrieved 04, April 2020
- Iron-functions-in-the-body, health: news-medical.net, Retrieved 25, August 2020
- Calcium-functions-food-sources-supplements-deficiencies-and-toxicity: netmeds.com, Retrieved 28, March 2020
- Types-of-vitamins-and-nutrients/iodine-a-critically-important-nutrient, vitamins-and-supplements, food: eatright.org, Retrieved 07, Jan 2020
- Manganese-functions-food-sources-deficiency-and-toxicity: netmeds.com, Retrieved 25, June 2020

5

Enzymes as Food Additives

Enzymes are proteins that can act as biological catalysts. They can be used as food additives for fruits and vegetables processing, wine processing, and baking and meat industry. This chapter closely examines these key applications of enzymes as food additives to provide an extensive understanding of the subject.

Enzymes are nontoxic protein substances that occur naturally in foods or may be produced by microorganisms or biotechnology to catalyze various reactions. They are early inactivated by a specific PH and temperature. Although the presence of some enzymes may produce negative quality changes, other enzymes are often intentionally added to foods for their beneficial effect. Microorganisms are responsible for producing some of the enzymes desired in foods; thus, the microbes may be intentionally added to food.

Some examples of enzymes that are additions to other foods include bromelain (from pineapples), ficin (from figs), and papain (from papaya). The enzymes act as meat tenderizers of muscle tissue or connective tissue. Amylases hydrolyze starch in flour and are used along with acids in the production of corn syrup. Invertase used to hydrolyze sucrose and prevent its crystallization. Pectinases clarify pectin-containing jellies or juices; proteases may be used a meat tenderizers, to create cheeses from milk (rennin), and to produce soy sauce. Glucose oxidase is added to foods such as egg whites in order to prevent the Maillard browning.

Enzymes Application for Fruits and Vegetables Processing

Enzymes are important throughout the life cycle of fruits and vegetables. Enzymes remain active even after harvest and may have undesirable effects on color, texture, taste, smell, and nutritional value. The use of enzymes aids in the processing of fruits and vegetables, acting mainly in the hydrolysis of polysaccharides and facilitating the extraction of intracellular compounds. As the content of these biomolecules differs in each plant in composition and quantity, different enzymes are required for each process. The enzymatic process has advantages compared to the use of chemicals, such as higher specificity, mild reaction conditions, and less waste production. Main enzymes used in fruit and vegetable processing include pectinesterase, pectinase, protopectinase, hemicellulase, α-amylase, glucoamylase, etc.

Pectinases

Pectins include protopectins, pectin acids, and pectic, and represent around 0.5% to 4.0% of the total weight of fresh fruits as well as constituents from 2% to 35% of the middle lamellae, which connects adjacent cells.The pectin can be degraded by pectin lytic enzymes, which may be from vegetable or microbial origin. These enzymes are obtained from fruits and even fermentation of by-products of juice extraction. The pectinesterase (polymethyl galacturonate esterase, PMGE, EC. 3.1.1.11) catalyzes the hydrolysis of methyl ester groups from pectin, releasing methanol and converting pectin in pectane. The addition of pectinase facilitates hydrolysis of carbohydrates that allows better squeezing of fruit, providing increased extraction yield and enhancing the concentration of acids, flavorings, and colorants. Consequently, pectinases play a vital role in the food processing industries, especially in the production of beverages, which are widely used for extraction of oils, syrups, and starches. These enzymes are used for reducing the texture of vegetables in order to facilitate processing. Frequently, combinations of pectinases, cellulases, and hemicellulases, known as macerating enzymes, are employed in juice production, in order to facilitate maceration, liquefaction, and clarification.

Cellulases

Cellulases promote the hydrolysis of cellulose, which is a polysaccharide composed of several glucose units. These enzymes are highly specific biocatalysts that act synergistically to release sugars, of which glucose has major industrial importance due to the possibility of its conversion into ethanol, sweeteners, phytohormones, and organic acids, among others. The main commercial preparations of cellulases are obtained from filamentous fungi. Cellulases are used in the beverage industry for the production of fruit juices and wine processing. They facilitate the extraction of juices and maceration for production of fruit nectars by breaking the cellulose chains that helps hold the liquid in the plant cells. In addition, it breaks unpleasant-tasting compounds, releasing flavoring substances and improving beverage aroma and flavor.

Xylanases (Hemicellulases)

The hemicellulose xylan is the most abundant biopolymer in the world, representing about 30% of the renewable and organic carbon. Hemicellulases (xylanases) are a group of enzymes that degrade hemicellulose polymers. These enzymes are produced by bacteria, fungi, algae, protozoa, nematodes, crustaceans, insects, and plants. Xylanases are used together with cellulases and pectinases for clarification and liquefaction in the fruit and juice industry.

α-Amylases

α-Amylase occurs in plants, mammalian tissues, and microorganisms. Fungi α-amylase is thermosensitive, and it is inactivated at temperatures higher than 70 °C. α-Amylases account for about 30% of the enzymes in the world market and are applied in many industrial processes, such as starch saccharification, the textile industry, paper, brewing, baking, detergent, cakes, fruit juices, starch syrups, distilleries, preparation of drugs for stomach problems, and other pharmaceuticals.

Processors wishing to produce clear concentrated fruit juices must overcome many challenges, including handling different kinds of fruits in the same factory, and managing fluctuating quantities of fruit with variable compositions and textures depending on their ripening stage. Dealing with these factors requires efficient and reliable tools, such as equipment and enzymes. Enzymes are processing aids used worldwide for fruit processing, particularly for the production of clear fruit juice and concentrate. They offer numerous advantages:

- Economic: Enzymes increase the overall productivity of the processing plant. The juice remains stable, long term, without additives or preservatives, storage volumes are reduced and shipping weight reduced.

- Quality: Fast juice processing with enzymes lowers the risk of microbial spoilage, reduces oxidation and improves juice and concentrates shelf life. Pectin hydrolysis of fruit cell walls weakens cells and vacuoles and thereby maximizes extraction of their components such as the red colour from berries (anthocyanins), aromas and antioxidants of phenolic type known for their positive effect on human health, particularly heart disease prevention.
- Sustainability: The use of enzymes has a positive effect on sustainable production. They lower energy consumption (electricity, steam and water), reduce waste flow by maximizing fruit use and reduce dependency on chemicals used in equipment cleaning products.

Vegetable juices are most commonly made from tomato, carrot and red beet. Vegetable Juice is one commercial example, made mainly from tomatoes and the juices of other additional vegetables like beets, celery, carrots, lettuce, parsley, watercress and spinach. Consumer demand for vegetable juices is increasing and fruit and vegetable juice combinations are becoming more popular. Numerous products came on the market. These new drinks are cloudy and pulpy. It may be easy to produce tomato juice without enzymes, but this is not the case for carrot, leek or cabbage. In fruit, pectin is the major component to break down in order to facilitate juice extraction. But vegetables also have a high content of dietary ligno-cellulosic fibres, which hinder juice extraction. Vegetable juice processing therefore requires more cellulases in addition to pectinases to reduce viscosity sufficiently for juice extraction using a decanter.

Rapidase is recommended for vegetable juice extraction. It contains pectinases and cellulases. After blanching (to limit oxidation), crushed vegetables are cooled to 50oC and mixed with enzymes. After 1–2 h, the juice is extracted using a decanter, pasteurized and possibly concentrated. The juice's carotene content is not decreased by enzymes, and the cloud is stable for the duration of the juice's shelf life.

New Developments in the Application of Enzymes in Fruit and Vegetable Processing

In recent years many new developments have been observed in the application of enzymes in fruit and vegetable processing.

Citrus Peeling

The first step in the preparation of citrus juices is the peeling of the fruits. This is a mechanical process requiring energy. Pectinases are used to soften the peel by disruption of the albedo and thereby facilitate a significant reduction in energy costs. Current industrial practice is starting with pectinase treatment of whole fruit, followed by a vacuum infusion treatment with a pectinase solution like Rapidase Intense and Peelzyme containing pectinesterase and polygalacturonase; thereafter the peel can be removed easily (1–2 hours) and the enzyme solution can be recycled.

Whole Fruits

Processing of whole fruit or fruit parts requires several precautions to safeguard the firmness of the fruits. Pectins consist of very complex structures giving strength to fruits but are sensitive to mechanical pressure (shear), heating (chemical hydrolysis), pasteurization, storage (polymer dehydration),and osmotic pressure. Moreover, most pectinase preparations consist of multiple enzymes leading to weakening of the pectin polymers.

For example, demethylation by pectin methyl esterase (PME) exposes the homogalacturonan backbone, which will be further degraded by enzymes like polygalacturonase (PG) and rhamnogalacturonase (RG), causing physical weakening of fruits. The Firm Fruit concept is based on the use of a PG and RG-free PME in combination with calcium, which binds to the freed pectic acid in situ to form insoluble calcium pectates. Other developments are the inhibition of PG by plant-born PG inhibitors thereby preventing PG activity in the fungal enzyme mixtures.

Immobilized Enzymes

Traditionally, enzymes in the fruit and vegetable processing industry are applied as liquids or powders, while enzymes in pharma are often immobilized. Although it adds an additional step (and thus costs) in the preparation of the enzyme, improved characteristics like a lower pH optimum and increased half-life (through recycling) can turn this into an attractive opportunity. First examples are shown for polygalacturonase and tannase.

Preventing Haze Formation

There are several ways to clarify extracted juices. In fruit juices, this is traditionally done with bentonite, silica gel, or gelatine followed by filtration. Although these methods do work, they remove a considerable amount of the antioxidant phenolics, which can form haze-causing interactions with the proteins present. The addition of gallic acid in combination with various proteases also reduced the haze formation drastically, but retained much more of the beneficial phenolics. The best-performing enzyme is Enzeco Fungal Acid Protease produced from A. niger, which reduces only 12% of the phenolics, while the traditional methods reduce up to 30%.

Preventing Cloud Loss

To produce cloudy juices, the process from extraction to pasteurization must be fast to minimize the effect of plant endogenous enzymes like pectinases and oxidases. Besides active inhibition of the plant enzymes by virtue of their inhibitors, α-amylases and gluco amylases can be added to reduce the starch content, as dissolved starch retrograde during cooled storage and can form precipitates.

Enzymes Application for Wine Processing

The production of wine is a biotechnological process marked with tradition. However, research into winemaking is providing an understanding of the biochemical reactions that occur in grape juice during its transformation into wine. Enzymes play a definitive role in the process of winemaking. Indeed, wine can be considered the product of enzymatic transformation of the grape juice. Enzyme activity is required as early as possible in the pre-fermentation stage. The enzymes originate not only from the grape itself, but also from yeast sand fungi and other microorganisms. Moreover, the action of such endogenous enzymes is nowadays reinforced and extended by the use of exogenous industrial enzymes. Enzymes coming from adventitious microorganisms infecting the grapes may also play a role. The applications of industrial enzymes are discussed below:

Applications of Enzymes

Enzyme preparations are used throughout the whole winemaking process:

- On the grapes (weakening, maceration).
- In the must and the press wine (clarification and sedimentation).
- In the young wine at the end of the fermentation.

Enzymes are added to the process in different ways. In all cases solutions are made of the enzyme (granular enzyme preparations are also readily soluble).

Actual addition proceeds via:

- Spraying the enzyme solution on the grapes.
- Dosing the solution via a pump system directly into the inlet of the press.
- Adding the solution to the tank before clarification.

Due to their specific character and action, enzymes release molecules that play a role in the taste characteristics of wine (colour, aroma, structure) or break down molecules that limit the effect and efficiency of technological treatments, like pressing, clarification and filtration. Better control of fermentation processes and enzyme production has lead to the possibility of purifying enzymes, for example, excluding the presence of cinnamyl esterase.

Simultaneous use of bentonite and enzymes should be avoided because the enzyme will be inhibited due to specific adsorption to the bentonite and the bentonite will become less effective due to blocking of the active sites by enzyme protein. Bentonite treatment should preferably take place after the enzyme treatment. Addition of bentonite will help in flocculation of enzymatically hydrolyzed pectins.

The activity and efficiency of an enzyme can vary enormously, depending on temperature and pH. Must pectinases can be used in the temperature range of 10–55 °C. Below 10 °C the enzyme dosage should be increased. Above 55 °Cthe enzyme will be inactivated. β-Glucanase can only be used above 15 °C and requires a longer incubation time. Enzyme dosages should be also increased at low pH values.

Enzymes are not inhibited by sulfur dioxide (SO_2) at levels that are acceptable in wine. Inhibition by polyphenols may occur in red wines. Increasing the enzyme dosage can counteract the effect. Alcohol up to a level of 14% (w/v) has no negative influence on enzyme action. In fact it has been reported that alcohol has an activating effect on β-glucanases used for aroma liberation.

Enzymes for Pressing and Maceration

Increased understanding of the way enzymes act on the polysaccharides of the grape cell walls, such as arabinans, galactans, arabinogalactans and arabinoxylans, makes it possible to optimize the usage of enzymes in winemaking.

The mechanism of action of pectolytic enzyme preparations, usually called pectinases. As far as is currently known, the enzyme activities involved in hydrolyzing pectic substances are pectin esterase, polygalacturonase, pectin lyase, rhamnogalacturonase, rhamnogalacturonan acetylesterase, arabinase and galactanase.

Other enzyme activities are of hemicellulase and cellulase type and are normally present in varying amounts in pectinase preparations. The combined action of all these enzymes leads to a partial hydrolysis and solubilization of acid and neutral polysaccharides present in the pectocellulose wall and middle lamella of the grape cells. These structures are responsible for the characteristic stiffness of the berries. The result of this hydrolysis is the release of juice, aromas and colour in a selective and controlled way. The remaining cell wall fragments are solubilized in the colloidal wine solution.

Enzyme Preparations for Aroma Liberation

The varietal character of (white) wines is especially defined by the presence of aromatic molecules, among which mono terpene alcohols play a well-defined role. These compounds are found in grapes as free, volatile, odourous molecules, and as non-volatile glycosidic precursors, also known as 'bound terpenes'. In many grape varieties the amount of bound terpenes can be higher than the amount of free aromatic terpenols. Consequently, the distinctive character of wines could be increased by releasing the glycosidically-bound terpenes.

Enzyme preparations containing glycosidase activity were developed to improve the aromatic profile of certain wines. These enzyme preparations are added at the end of the alcoholic fermentation and during 'retapping' of wines (pouring into another vessel), which are not treated with bentonite, in order to prevent enzyme inhibition. The optimal temperature for this enzyme treatment should be above 15 °C and an incubation time of several weeks to one month is required. The aroma development has to be controlled by tasting, and the enzyme action is finally inhibited by addition of bentonite. Small amounts of this compound (20 g/hl) are usually sufficient to block all enzyme activity.

Since the mechanism of hydrolysis of terpenyl glycosides has been elucidated, various different fungal enzyme preparations have been developed capable of enhancing wine aroma. Aspergillus spp. Pectinase and glucanase preparations have good results on wine making conditions. These enzymes can be used in fermented juice as soon as glucose has been consumed by the yeast, or in young wines. In this way the variety in aroma and bouquet of certain wines can be improved.

Enzymes for Colour Extraction

The extraction of phenolic compounds generally occurs with the maceration of the mash during alcoholic fermentation and depends on the variety and quality of the grapes and on technological parameters such as crashing, maceration time, temperature and pumping. Even though colour extraction trials are difficult to conduct because of the heterogeneity of the mash and the influence of processing factors, good results have been obtained on an industrial scale using enzymatic treatments. Improved anthocyanin extraction was obtained by enzyme addition at the beginning of maceration. However, more fundamental research is required to understand all mechanisms involved in colour extraction fully.

Other Enzymes Used in Winemaking

Urease

An enzyme preparation containing the enzyme urease was developed in Japan. Urease catalyses the hydrolysis of urea to carbon dioxide and ammonia. This limits the formation of ethyl carbamate. This molecule, suspected to be carcinogenic and mutagenic at high dose rates, is predominantly urea. The enzyme is produced by Lactobacillus fermentum.

Lysozyme

Microbial stabilization of wines is accomplished by the addition of sulfur dioxide. Stabilization is guaranteed by lowering the sulfur dioxide level in combination with treatment with lysozyme. Lysozyme is derived from egg white protein and is used in the pharmaceutical and food industries. The enzyme has the ability to break down the cell walls of lactic bacteria and its activity increases with increasing pH, Lysozyme treatment is recommended for:

- Blocking the malolactic fermentation in white wines.
- Stabilizing red wines after fermentation.
- Treatment of problematic end stages of alcoholic fermentation.
- Preventing premature malolactic fermentation in wines.

New Developments

Recent investigations have stimulated interest in another group of enzymes, oxidases. In particular, a polyphenol oxidase originating from the fungus Myceliophtera thermophile and produced by Aspergillus oryzae has been developed for application in the wine industry.

Cork Cleaning

This application deals with the corks used for closing wine bottles. The enzyme removes phenolic compounds (which can give an off-flavour to the wines) and is applied during cleaning of the corks. This enzyme treatment significantly reduces the amount of water-extractable total polyphenols in comparison with crude, unwashed corks. It is comparable with chlorine washing, but has the advantage that the level of trichloro anisol (TCA) is reduced instead of increased, as is the case with chlorine treatment.

Colour Stabilization

Removal of phenolic compounds to prevent oxidation, a haze and flavour change is usually undertaken with fining agents such as gelatin or polyvinyl pyrrolidone. Treatment with enzymes such as laccases, tannases and peroxidases has been investigated for juice or wine stabilization. Fungal laccase has ability to react with a wide range of phenolic compounds. Laccase treatment can enhance the effect of conventional fining treatments.

Enzymes Application in Baking and Meat Industry

Enzymes Application in Baking

Baking is a common name for the production of baked goods, such as bread, cake, biscuits, crackers, cookies, tortillas etc. Enzymes are rapidly becoming very important to the baking industry. They are used in baking to make consistently high-quality products by enabling better dough handling, providing anti-staling properties and allowing control over crumb texture, colour, taste, moisture and volume.

In baking industry enzyme is mostly applied to bread making. Most enzymes applied in bread making can also be applied in the other baking applications. Depending upon the raw materials used in bakery products, amylases, hemicellulases, lipases, oxidases, cross-linking enzymes and proteases may be applied to improve the quality or modify the texture of the baked products.

Bakery products have undergone radical improvements in quality in terms of flavour, texture and shelf life. The usage of enzymes is the biggest contributor to these improvements. Baking enzymes are used as flour additives and they are used in dough conditioners to replace chemical ingredients and to perform other functions. The baking industry predominantly makes use of different types of enzymes. Amylases are used to convert starch to sugar and to produce dextrins. For strengthening and bleaching of the dough, oxidases are used. Hemicellulases and proteases are the enzymes which have an effect on wheat gluten. While hemicellulases improve gluten strength, proteases reduce gluten elasticity. All these enzymes together play an important role in maintaining volume, crumb softness, crust crispiness, crust colouring or browning and in maintaining freshness.

Application of Enzymes in Bread Making

Bread is the product of baking a mixture of flour, water, salt, yeast and other ingredients. The basic process involves mixing of ingredients until the flour is converted into dough, followed by baking the dough into bread. The aims of the bread-making processes are to produce dough that will rise easily and have properties required to make good bread for the consumer. To make good bread, dough made by any process must be extensible enough to expand during proofing. Bread dough must also be elastic. Elastic dough has the strength to hold the gases produced while rising and stable enough to hold its shape and cell structure.

For decades enzymes such as α-amylases have been used for bread making. Due to the changes in the baking industry and the demand for more varied and natural products, enzymes have gained more and more importance in bread formulations. Through new and rapid developments in biotechnology, a number of new enzymes have recently been made available to the baking industry. One example is pure xylanase, which improves the dough machinability. A lipase has gluten strengthening effect that results in more stable dough and improved crumb structure similar to

DATEM or SSL/CSL and α -amylase that has a unique anti-staling effect. There are certain enzyme discussed below which are applicable to bread and other baked products.

Amylases

α-amylases are the enzymes which are most frequently used in bakeries. The reasons for this are their positive influence on bread volume, crumb grain, crust and crumb colour, flavour development and anti-staling effect. Amylases also have a positive effect on dough development.

α-Amylases are endoglucanases. This means that they hydrolyze random α-1,4 and α-1,6 linkages. Amylases can act on damaged or gelatinized starch, since these are susceptible to enzymatic attack. Suitable dosages of amylase lead to the desired improvement of dough and the final product. However, extensive degradation of damaged starch due to too high levels of α-amylase leads to sticky dough. The volume and crumb structure improve with increasing levels of amylase added to the flour. Amylases also have effect on staleness of bread. These enzymes have a significant action on gelatinized amorphous starch. Modification of gelatinized starch results in a clear anti-staling effect.

Xylanases

Xylanases are broadly used in bread making and depending on the application there is generally an appropriate xylanase or a mixture of different xylanases that gives the desired effects in terms of dough-handling properties; stability, oven spring and volume. Xylanase improves the dough machinability.

Lipases

Specific lipases are claimed to improve dough-handling properties, dough strength and stability, dough machinability and oven spring. Besides this, lipases also improve crumb structure and crumb whiteness. Use of lipases in baking is claimed to be alternatives to chemical dough strengtheners and emulsifiers. However, the technical and commercial benefits are limited. Like DATEM, lipases also increase surface pressure of gas cells which leads to a better distribution of more stable and smaller gas cells. This results in finer, silkier crumb structure with optically whiter colour, better dough-handling properties and, to a certain extent, a larger loaf volume.

Oxidases

In bread making, after mechanical development of the gluten network, the three-dimensional protein structure needs to be stabilized by oxidants. Small amounts of oxidizing reagents, such as potassium bromate or dehydro ascorbic acid, improve the dough handling and baking characteristics of wheat flour which increases loaf volume and improves bread crumb. Increasing demands by consumers for more natural products with fewer chemicals, and especially concerns about the possible risks of bromate in food, have created the need for bromate replacements. Therefore, oxidases are gaining increasing attention in the baking industry. Glucose oxidase for bread making has been known very well. Glucose oxidase has good oxidising effects that result in a stronger dough. It can be used to replace oxidants such as bromate and ascorbic acid in some baking formulations and procedures. In other formulations it is an excellent dough strengthener along with ascorbic acid.

Proteases

Proteases have a long history in bread making and are traditionally used to treat 'bucky' dough resulting from overly strong and too elastic flours. Originally the aim of protease addition is to improve softness, dough-handling properties and dough machinability. However, proteases have more functional effects. Functional effects of proteolytic enzymes are reduction of mixing time; improvement of dough machinability; improvement of gas retention due to better extensibility; improve pan flow in bun and roll production; improvement of crumb texture; improve water absorption; improve colour and flavour.

Application of Enzymes in Cake and Muffin Production

Cakes and muffins are produced from the basic ingredients like wheat flour, sugar, eggs and fat source. Cakes are produced by mixing the constituents into a liquid batter and include air to form a foam. The air expands during baking and the foam transforms into a sponge because of the viscosity increase caused by the gelatinization of starch. A cake or muffin batter may be considered as an oil-in-water emulsion which can be stabilized by emulsifiers.

Emulsifiers are important part of cake and muffin recipes. They are added to support the incorporation of air and to improve dispersion of fat in the batter and will also stabilize expanding gas bubbles in the batter during baking. Replacement of these emulsifiers by a commercial lipase in the production of cake is possible. Addition of commercial lipase reduces the surface tension and surface viscosity at the air/water interface of batter. This indicates that surfactants are created which replaces proteins at the air/water interface. After baking, it results in increase in cake specific volume and maintaining a fine crumb structure. It also improves Eating quality and freshness.

In a cake recipe, when the amount of egg is reduces, the quality of the cake in general will deteriorate. This deterioration can be counteracting by adding phospholipase to the cake batter. Phospholipases increase cake volume and improve cake properties during storage like increase cohesiveness, springiness and elasticity. Apart from lipolytic enzymes, starch-degrading enzymes can also apply in cake production. Starch-degrading enzymes prevent cake staling.

Bacterial amylase can use as part of a cake powder conditioner, which can improve the quality of cake in general and more specifically the softness of the crumb and the shelf life of the product. Amylomaltase or glucosyltransferase is a thermostable enzyme able to hydrolyze oligosaccharides from amylose.

Proteases are used to lower the viscosity of cereal flour suspensions and to avoid checking during biscuit baking. Nowadays it also apply to retard staling of the cake crumb. Use of intermediate thermostable alkaline proteases, for example keratinase and thermitase, which has no perceivable influence on the dough rheology, has a pronounce effect on the softness and retardation of the crumb hardness, resulting in a prolonged shelf life. Specific proteases may also be applied to improve the flavour of cakes.

Application of Enzymes in Biscuit, Cookie and Cracker Production

The principal ingredients used in the manufacture of biscuits and cookies are wheat flour, fat and sugar. Water plays an important role in the biscuit-making process but is largely removed during baking. Biscuit manufacturing generally includes several steps like mixing, resting time, machining and finally baking. Crackers belong to the category of hard dough biscuits and can be either chemically leavened or fermented. Sodium metabisulphite (SMS) is currently used in the baking industry to soften biscuit dough. In particular, SMS is used in the industry to reduce shrinking of dough pieces and irregular sizing of baked products.

The use of protease to modify gluten quality has been very well known. Protease can be used in cracker in order to increase dough extensibility. This allows cracker manufacturers tight control over dough consistency. Compared to sulphite, proteases work in a different way since they hydrolyze the inner peptide linkages of gluten proteins, whereas SMS increases extensibility by breaking the disulphide bonds. The texture of the biscuits obtained will also be more open and tender.

Use of an oxidation-sensitive protease, such as papain, in combination with an oxidizing enzyme (such as glucose oxidase) producing an oxidizing agent, can enable biscuit manufacturers to mimic the effect of sulphite in dough. Sometimes, papain hydrolyzes the gluten to such a degree that the resulting dough is not suitable for biscuit baking. The combination of papain and glucose oxidase results in a quick decrease in dough consistency to a desired level. This level remains more or less constant over time. The overall results show that glucose oxidase was able to reduce the action of papain over time. Papain is an interesting protease to use since it has a strong hydrolytic action on glutenins but can be spontaneously stopped by natural oxidation of the dough. Possibly, addition of glucose oxidase generates hydrogen peroxide from glucose present in the dough which inactivates papain in an irreversible way. This allows the use of papain in the dough with quite a great security and without fearing any adverse process effects.

More even baking can also be achieved by improving the properties of the dough through the addition of hemicellulose and cellulose-degrading enzymes. It has been found that the enzymes make the dough softer, requiring less water, less energy input, finally resulting in increased factory output. The use of hemicellulases in cracker dough can potentially be also very useful. The partial breakdown of the water-extractable hemicellulose fractionwill lowers the water binding capacity. Hence, more water will be available and softer dough will be achieved. Consequently less water is needed to prepare the dough. Furthermore, a reduction in baking time will be achieved, as well as an improvement in quality through more even baking resulting in reduced checking.

α-Amylases only play a minor role in biscuit manufacturing. Due to the fact that α-amylase is able to produce dextrins from damaged starch, they will play a role in the enzymatic browning during baking, resulting in darker biscuits. Also, the addition of (fungal) α-amylase will potentially prevent checking as well as creating a leavening effect and improved flavour development. The amylase will act on the damaged starch granules, thereby providing food for yeast to generate carbon dioxide, while at the same time liberating water from the damaged starch. This will improve the distribution of water throughout the dough, creating more uniformity, hence less problems with checking after baking.

Use of a pentosanase will reduce checking in crackers by reduction of the water content and will be particularly useful in low fat and/or high-fibre formulations. Dough containing low level of fat or high level of fibre do require higher amount of water to be added to the process in order to achieve good machinability. This water also needs to be removed during baking, resulting in longer baking times. The addition of hemicellulases will result in lower water binding capacity; hence more water will be available for easier processing.

Use of Enzymes in Tortilla

Tortillas are unleavened, flat, round breads made from wheat or corn. Flour tortillas are made of wheat flour, water, shortening, and salt, preservatives, leavening agents, reducing agents and emulsifiers.

Staling of tortillas involves the starch in the amorphous phase and does not significantly interfere with the amylopectin crystallization. It is proposed that bacterial α-amylase partially hydrolyzes the amylose, starch bridging the crystalline region and protruding amylopectin branches. Starch hydrolysis decreases the rigid structure and plasticity of starch polymers during storage. The flexibility of tortillas results from the combined functionalities of the amylose gel and amylopectin solidifying the starch granule during storage.

Enzymes Application in Meat Industry

In the meat industry, there are two distinct applications in which enzymes can markedly boost the manufacturing process and upgrade meats of poorer quality. These applications are tenderization of too tough meat parts and restructuring of fresh low-value meat pieces and trimmings to higher quality steaks. In meat industry predominantly protein-degrading enzymes have been used. Cross-linking enzymes like transglutaminases have been used as texture improvers. In addition to these, novel enzymes are discovered. Structure engineering by oxidative enzymes and flavour design by lipases, glutaminases, proteases and peptidases are examples of emerging enzyme technologies in the food sector. Consumers' demands for high quality and moderate price meat products have been the driving forces to develop enzymatic methods to add value to lower quality meat piece.

Proteases have an important role in meat processing in tenderization. Papain, bromelain and ficin are being used at commercial scale meat tenderization. Furthermore, proteases have been used for bone cleaning and flavour formation in the meat industry. Lipases can be used for flavour formation in sausage production. Transglutaminase can be used for structure engineering for tailoring the structural properties of different processed and heated meat products. Oxidative enzymes can be an alternative for Transglutaminases to generate cross-links in protein matrices. Oxidoreductases including tyrosinases and laccases are reported to cross-link meat proteins. l-Glutaminase (l-glutamine aminohydrolase) produced by starter cultures has an important role in flavour formation, for example in sausage production.

Meat Tenderization with Added Enzymes

Texture and tenderness are the most important attributes in meat products. Methods to increase tenderness include natural aging, electric stimulation, mechanical blade tenderization and use of added proteolytic enzymes. The most widely used exogenous enzymes in meat tenderization are the plant enzymes papain, bromelain and ficin.

The way of applying tenderizing enzymes in meat industry depends on the actual target. If the aging time of high-grade meat parts has to be shortened, the main action of protein hydrolysis should be on myofibrillar proteins. If the tenderness improvement of lower grade meat cuts or meat from the connective tissue proteins, mostly collagen, should be the target of proteolysis. The methods and challenges to tenderize meat sold raw to consumers differ from those needed for cooked meat.

Plant proteases mainly used to tenderize meat act unfortunately more actively on other meat proteins than on collagen. Therefore attempts to tenderize collagen-rich connective tissue inevitably led to too extensive hydrolysis of non-collagen proteins, resulting in too soft (mushy) meat. To tenderize meat pieces with a high connective tissue content it is evident that an enzyme having pronounced activity against connective tissue but limited activity against myofibrillar proteins should be used.

Enzymatic Generation of Flavour in Meat Products

Flavour has a major role, along with tenderness, in acceptability of meat by consumers. The flavour of raw meat is quite bland. However, it contains non-volatile constituents that are essential flavour precursors which during processing and storage affect the taste of the meat product. In general, the flavour of processed meat is a result of either enzymatic action or chemical reactions such as pyrolysis of amino acids and peptides, sugar degradation, degradation of ribonucleotides, Maillard reactions, thiamine degradation and degradation of lipids. The main enzymatic reactions affecting meat flavour or formation of flavour precursors are proteolysis and lipolysis. Both groups of reactions are due to the contribution of endogenous proteases and lipases, enzymes of microbial origin naturally present in the product or enzymes added during the manufacturing process.

Proteolysis and Lipolysis in Meat Flavour Development

Dry-cured meat products are appreciated for their unique flavour. The compounds implicated in flavour generation arise from many sources, such as spices, sugar metabolism, lipolysis and lipid oxidation, proteolysis and amino acid degradation. During ripening proteolysis takes place, yielding for example polypeptides, peptides and free amino acids, which are involved in taste and flavour development of meat products. Meat protein hydrolysis is mainly catalyzed by endogenous enzymes, such as cathepsins and trypsin-like peptidases as well as proteases produced by microorganisms are involved in the ripening process. These enzymes are mainly originating from Micrococcaceae but also from moulds and yeasts in those dry sausages in which they are present. Glutaminase addition to protease has an important role from the point of view of sausage production, especially regarding the deamidation of glutamine, since hydrolysis of the glutamine amide group produces ammonia, an acidity neutralizer and umami flavour. Umami can be described as savoury or broth-like taste with ability to enhance other flavours. Lipolysis constitutes another important group of enzymatic reactions which are related to aroma formation of fermented sausages. Phospholipases and lipases hydrolyze phospholipids and triacylglycerols forming free fatty acids. Unsaturated fatty acids are further oxidized to aroma volatile compounds. This oxidation may lead to the formation of aliphatic hydrocarbons, alcohols, aldehydes and ketones. Further alcohols react with free fatty acids forming some esters.

Effect of Enzymes on Ripening of Dry-cured Meat Products

Dry-cured meat products need a long period of ripening in order to allow the transformation of free amino acids and fatty acids through microbial (oxidative deaminations, decarboxylations) and/or chemical (Maillard reactions) ways to yield aromatic compounds (aldehydes, ketones, lactones, alcohols and esters). Because long ripening time involves a high cost of storage until a suitable matured state is reached, many attempts have been made to shorten this period. Proteases and lipases have been used for this purpose. However, it has been found that addition of proteinases and lipases alone is not useful in shortening the ripening time. This is because the final flavour also depends on subsequent generation of volatile compounds through lipid oxidation and amino acid catabolism. Therefore, to shorten the ripening of sausages, it is necessary to create conditions, for example by adding an efficient starter or by adding other types of enzymes, which stimulate formation of volatiles in a shorter time than usual.

Structure Engineering by Cross-Linking Enzymes

Apart from the hydrolytic enzymes affecting the tenderness of meat or generation of flavour, the functional properties of meat proteins can be modified by cross-linking enzymes. These enzymes are used to bind fresh meat pieces together and to tailor the structural properties of various processed meat products. The main target protein in meat for cross-linking enzymes is the myofibrillar protein myosin. Cross-linking enzymes are generally capable of positively affecting gelation and consequently the texture of meat gels. Transglutaminase has been the main cross-linking enzyme which is applied industrially in meat protein modification.

Restructuring of Unheated Meat

High-quality meat products at moderate prices demanded by consumers have been the driving forces to develop methods to restructure low-value cuts of poorer quality to improve their market value by making them palatable steaks resembling intact muscle and to maximize the efficiency of carcass utilization. Traditionally salt and phosphates in conjunction with heat treatment have been used to bind meat pieces together. Unheated comminuted products are usually frozen to enhance binding. Nowadays, when consumers demand fresh, unfrozen meat as well as lower salt contents, technologies have been developed to eliminate the need for freezing and to enable the use of less salt. One of these technologies is the enzyme-aided restructuring, which has been used on a commercial scale for some time and is still the main Transglutaminase application in the meat sector. Transglutaminase has been found to improve the strength of restructured meat protein gels with or without added salt and phosphates.

Processed Meat Systems

In addition to binding fresh meat pieces together, the effects of Transglutaminase have been used in isolated meat protein systems and model meat products aiming at improved textural properties. Transglutaminase-catalyzed formation of additional covalent bonds in structural meat proteins leads to firmer gel structures.

Other Applications

Proteolytic enzymes are also potential tools for valorization of different meat by-products. Protein hydrolyzates with strong meat flavour can be used in soups, sauces and in ready meals. Proteases can be applied for production of protein hydrolyzates from different meat by-products such as bones, sheep visceral mass, chicken by-products or bovine byproducts. These hydrolyzates can be used as flavour enhancers, as seasoning additives, as nutritional additives to low-protein food products or as animal feed supplements when not suitable for food use. Optimization of the type proteolytic enzymes used in the treatments is needed in order to avoid formation of bitter hydrolysis products when food applications are targeted. The flavour intensity depends on the free amino acid content and on the type of peptides present and their reactions during the process, and therefore the presence of endoproteases and exo-peptidases has to be optimal. Enzymes can also be used for treatment of fresh bones to be suitable for gelatine production. This process produces meat extract and cleaned bones for subsequent gelatin manufacture.

Permissions

All chapters in this book are published with permission under the Creative Commons Attribution Share Alike License or equivalent. Every chapter published in this book has been scrutinized by our experts. Their significance has been extensively debated. The topics covered herein carry significant information for a comprehensive understanding. They may even be implemented as practical applications or may be referred to as a beginning point for further studies.

We would like to thank the editorial team for lending their expertise to make the book truly unique. They have played a crucial role in the development of this book. Without their invaluable contributions this book wouldn't have been possible. They have made vital efforts to compile up to date information on the varied aspects of this subject to make this book a valuable addition to the collection of many professionals and students.

This book was conceptualized with the vision of imparting up-to-date and integrated information in this field. To ensure the same, a matchless editorial board was set up. Every individual on the board went through rigorous rounds of assessment to prove their worth. After which they invested a large part of their time researching and compiling the most relevant data for our readers.

The editorial board has been involved in producing this book since its inception. They have spent rigorous hours researching and exploring the diverse topics which have resulted in the successful publishing of this book. They have passed on their knowledge of decades through this book. To expedite this challenging task, the publisher supported the team at every step. A small team of assistant editors was also appointed to further simplify the editing procedure and attain best results for the readers.

Apart from the editorial board, the designing team has also invested a significant amount of their time in understanding the subject and creating the most relevant covers. They scrutinized every image to scout for the most suitable representation of the subject and create an appropriate cover for the book.

The publishing team has been an ardent support to the editorial, designing and production team. Their endless efforts to recruit the best for this project, has resulted in the accomplishment of this book. They are a veteran in the field of academics and their pool of knowledge is as vast as their experience in printing. Their expertise and guidance has proved useful at every step. Their uncompromising quality standards have made this book an exceptional effort. Their encouragement from time to time has been an inspiration for everyone.

The publisher and the editorial board hope that this book will prove to be a valuable piece of knowledge for students, practitioners and scholars across the globe.

Index

A

Alkaline Compounds, 10, 115

Artificial Flavourings, 1, 12

Artificial Sweeteners, 20, 47, 70, 73, 78

Ascorbic Acid, 4-5, 138, 150-151, 194 202-203, 229

Aspartame, 13, 19, 73-75, 77-78, 177

B

Benzoic Acid, 4

Bioemulsifiers, 110-114, 116

Biosynthesis Pathway, 66-68

Blanching, 151, 223

C

Carotenoids, 11, 26, 29-30, 32, 34 200-201

Chlorophyll, 27, 29, 31-32, 34-35

Choline, 206

Collagen, 126-127, 129, 208, 232-233

Crystal Growth, 108, 139, 171

Crystallization, 31, 71, 92-93, 107, 144 221, 232

D

Decarboxylation, 18, 60, 62, 67, 69, 203

Defoaming Agent, 93

Deoxyribonucleic Acid, 206

Dietary Fibre, 3

Diglycerides, 7-9, 35, 92, 95, 145

E

Electrolyte, 139, 188, 190-191

Emulsifiers, 3, 7-8, 16, 35, 54, 91-96 98-103, 107, 109-111, 115, 117 124-125, 137, 144-145, 147, 149, 170 208, 229-230, 232

Emulsifying Salts, 117-123

Enzyme Application, 221, 224, 228, 232

Enzyme Preparations, 225-226

Etherification, 153, 155, 158, 162-164 166-167, 169

F

Fats And Oils, 20, 30, 91, 96, 98, 107 109-110

Fatty Acid, 3, 8, 29, 92, 94, 97, 106, 110 113, 139, 145, 158

Flavour Biosynthesis, 59, 61, 64-66

Flavour Creation, 53, 55

Foaming Agent, 6-7, 9, 102

Folic Acid, 24, 206

Food Additives, 1-2, 10, 12-23, 31-32 39, 58, 70, 73, 77, 117, 124, 144-145 151, 171, 183, 221

Food Color, 29, 31

Food Safety, 14-15, 17-18, 20-22, 117

G

Gelling Agents, 3, 140, 167, 171, 173 175, 177-180

Glucose Oxidase, 221, 229, 231

Glutamine, 61, 232-233

Glycine, 64-65, 127

H

Humectant, 9, 93

Hydrocolloids, 57, 91, 126, 171-178, 180

Hydrophobic Bonds, 148

Hypersensitivity, 17-18, 24, 42

I

Intrinsic Viscosity, 147, 174-175

J

Junction Zones, 148, 174-176, 181

L

Lactose Intolerance, 9, 217

Leucine, 62-63

Linoleic Acid, 3

M

Maillard Reaction, 50

Melanin, 5

Melting Point, 8, 31, 73, 107, 132, 179

Methionine, 62, 127, 206, 220

Microbial Growth, 65-66, 150

Microorganisms, 4-5, 16, 35, 54, 64, 69 76, 82-84, 111-112, 116, 124, 126, 137 200, 221-222, 225, 233

Modified Starch, 145, 153-154, 157, 167 172-173, 175, 178

Monoglycerides, 6, 35, 92, 95, 97-98 107, 109-110

Myosin, 234

N

Natural Colorants, 10-11, 27, 34-37

Non-nutritive Sweeteners, 13, 73, 77-78

Nucleic Acids, 188, 203, 206

Nutritional Additives, 2-3, 183-184, 234

O
Oligosaccharides, 136, 141, 215-216, 230

P
Pantothenic Acid, 205
Pasteurization, 74, 123, 129, 224
Pectins, 141, 143-144, 147, 149, 179, 221 224-225
Phenylalanine, 19, 63-64, 66, 74, 84-85
Polyglycerol, 91-92, 94, 97
Polyoxyethylene, 7-8, 92
Polysaccharides, 3, 115, 130, 143, 149 154, 171-172, 174-177, 216, 221, 226
Proofing, 228
Protein Powders, 209-210, 213, 218-220

R
Riboflavin, 27, 29, 33-35, 204
Ribonucleic Acid, 206
Risk Management, 22

S
Saccharin, 13, 73-79
Sensory Analysis, 138
Sorbitol, 8, 13, 70-72, 78, 91, 150
Soy Protein, 207-208, 214-219
Spray Drying, 123
Stabilisers, 124-125
Stearic Acid, 7
Surfactant, 7-8, 93, 110, 145, 147
Synthetic Compounds, 5, 12-13

T
Thermal Processing, 144, 205
Tricarboxylic Acid Cycle, 64
Tyrosine, 5, 18, 63-64

U
Unsaturated Fatty Acids, 5, 233

V
Valine, 62

W
Whey Protein, 70, 120, 148, 208, 210-214